Zooarchaeology in the Neotropics

Mariana Mondini • A. Sebastián Muñoz
Pablo M. Fernández
Editors

Zooarchaeology in the Neotropics

Environmental Diversity and Human-Animal Interactions

Editors
Mariana Mondini
Laboratorio de Zooarqueología y
   Tafonomía de Zonas Áridas (LaZTA)
IDACOR, CONICET/Universidad
   Nacional de Córdoba
Córdoba, Argentina

Facultad de Filosofía y Letras
Universidad de Buenos Aires
Ciudad Autónoma de Buenos Aires, Argentina

Pablo M. Fernández
Instituto Nacional de Antropología y
   Pensamiento Latinoamericano—CONICET
Ciudad Autónoma de Buenos Aires, Argentina

Facultad de Filosofía y Letras
Universidad de Buenos Aires
Ciudad Autónoma de Buenos Aires, Argentina

A. Sebastián Muñoz
Laboratorio de Zooarqueología y Tafonomía de
   Zonas Áridas (LaZTA)
IDACOR, CONICET/Universidad
   Nacional de Córdoba
Córdoba, Argentina

ISBN 978-3-319-86134-0   ISBN 978-3-319-57328-1   (eBook)
DOI 10.1007/978-3-319-57328-1

Printed on acid-free paper

This Springer imprint is published by Springer Nature
The registered company is Springer International Publishing AG
The registered company address is: Gewerbestrasse 11, 6330 Cham, Switzerland

Contents

1 Zooarchaeology in the Neotropics: An Introduction 1
Mariana Mondini, A. Sebastián Muñoz, and Pablo M. Fernández

**2 Pinniped Capture and Processing: A Comparative Analysis
from Beagle Channel (Tierra del Fuego, Argentina)** 7
María Paz Martinoli and Martín Vázquez

**3 Use of Marine Fauna and Tool Stones in the South of Buenos Aires
Province (Argentina) During the Middle and Late Holocene** 25
Romina Frontini and Cristina Bayón

**4 Shell Mounds of the Southeast Coast of Brazil: Recovering
Information on Past Malacological Biodiversity** 47
Edson Pereira Silva, Sara Christina Pádua,
Rosa Cristina Corrêa Luz Souza, and Michelle Rezende Duarte

**5 Faunal Subsistence Resources in the Cañada Honda Locality
(Northeastern Buenos Aires Province, Argentina)** 61
Paula D. Escosteguy and Mónica C. Salemme

**6 Space Use Patterns and Resource Exploitation of Shell Middens
from the Río de La Plata Coast (*ca.* 6000–2000 Years BP),
Uruguay** 81
Laura Beovide, Sergio Martínez, and Walter Norbis

**7 Use of Animals During the Mid-Archaic and the Initial Period
in Pernil Alto: A Site in the Palpa Valleys, Southern Coast
of Peru** 103
Carmen Rosa Cardoza, Johny Isla, Markus Reindel, Enrique Angulo,
Hermann Gorbahn, and Lucía Watson Jiménez

**8 Taphonomy of Surface Archaeological Bone Assemblages
in Coastal Patagonia: A Case Study** 123
A. Sebastián Muñoz

**9   The Fossorial Faunal Record at the Beltrán Onofre Banegas-Lami
      Hernandez Archaeological Site (Santiago del Estero Province,
      Argentina): A Taphonomic Approach** 137
Luis Manuel del Papa, Luciano De Santis, and José Togo

**10   Archaeological Collagen Fingerprinting in the Neotropics;
      Protein Survival in 6000 Year Old Dwarf Deer Remains
      from Pedro González Island, Pearl Islands, Panama** 157
Michael Buckley, Richard G. Cooke, María Fernanda Martínez,
Fernando Bustamante, Máximo Jiménez, Alexandra Lara,
and Juan Guillermo Martín

**11   Osteometrics of South-Central Andean Wild Camelids:
      New Standards** 177
Mariana Mondini and A. Sebastián Muñoz

List of Contributors, Editors, and Peer-Reviewers

Contributors

Enrique Angulo Lima, Peru

Cristina Bayón Dpto. de Humanidades, Universidad Nacional del Sur, Bahía Blanca, Argentina

Laura Beovide Centro de Investigación Regional Arqueológica y Territorial-DICYT-MEC, Montevideo, Uruguay

Michael Buckley Manchester Institute of Biotechnology, Faculty of Life Sciences, The University of Manchester, Manchester, UK

Fernando Bustamante Universidad de Antioquia, Medellín, Colombia

Carmen Rosa Cardoza Lima, Peru

Richard G. Cooke Smithsonian Tropical Research Institute, Balboa, Ancón, Republic of Panama

Luciano De Santis Cátedra de Anatomía Comparada, Facultad de Ciencias Naturales y Museo, Universidad Nacional de La Plata, CONICET, La Plata, Argentina

Luis Manuel del Papa Cátedra de Anatomía Comparada, Facultad de Ciencias Naturales y Museo, Universidad Nacional de La Plata, CONICET, La Plata, Argentina

Michelle Rezende Duarte Laboratório de Genética Marinha e Evolução, Depto. de Biologia Marinha, Instituto de Biologia, Universidade Federal Fluminense, Rio de Janeiro, Brazil

Paula D. Escosteguy Instituto de Arqueología, FFyL, Universidad de Buenos Aires-CONICET, Buenos Aires, Argentina

Romina Frontini CONICET—Dpto. de Humanidades, Universidad Nacional del Sur, Bahía Blanca, Argentina

Hermann Gorbahn University of Kiel, Kiel, Germany

Johny Isla Andes: Centro de Investigación para la Arqueología y el Desarrollo, Lima, Peru

Máximo Jiménez Smithsonian Tropical Research Institute, Balboa, Ancón, Republic of Panama

Alexandra Lara Smithsonian Tropical Research Institute, Balboa, Ancón, Republic of Panama

Juan Guillermo Martín Universidad del Norte, Barranquilla, Colombia

María Fernanda Martínez Smithsonian Tropical Research Institute, Balboa, Ancón, Republic of Panama

Sergio Martínez Depto. de Evolución de Cuencas, Facultad de Ciencias, Universidad de la República, Montevideo, Uruguay

María Paz Martinoli Centro Austral de Investigaciones Científicas (CADIC), CONICET, Ushuaia, Argentina

Mariana Mondini Laboratorio de Zooarqueología y Tafonomía de Zonas Áridas (LaZTA), IDACOR, CONICET/Universidad Nacional de Córdoba, Córdoba, Argentina

Facultad de Filosofía y Letras, Universidad de Buenos Aires, Ciudad Autónoma de Buenos Aires, Buenos Aires, Argentina

A. Sebastián Muñoz Laboratorio de Zooarqueología y Tafonomía de Zonas Áridas (LaZTA), IDACOR, CONICET/Universidad Nacional de Córdoba, Córdoba, Argentina

Walter Norbis Instituto de Biología, Depto. de Biología Animal, Facultad de Ciencias, Universidad de la República, Montevideo, Uruguay

Sara Christina Pádua Laboratório de Genética Marinha e Evolução, Depto. de Biologia Marinha, Instituto de Biologia, Universidade Federal Fluminense, Rio de Janeiro, Brazil

Markus Reindel Deutschalnd Archaeologist Institut, Bonn, Germany

Mónica C. Salemme Centro Austral de Investigaciones Científicas—CONICET and Universidad Nacional de Tierra del Fuego, Ushuaia, Argentina

Edson Pereira Silva Laboratório de Genética Marinha e Evolução, Depto. de Biologia Marinha, Instituto de Biologia, Universidade Federal Fluminense, Rio de Janeiro, Brazil

Rosa Cristina Corrêa Luz Souza Laboratório de Genética Marinha e Evolução, Depto. de Biologia Marinha, Instituto de Biologia, Universidade Federal Fluminense, Rio de Janeiro, Brazil

José Togo Facultad de Humanidades, Ciencias Sociales y de la Salud, Universidad Nacional de Santiago del Estero, Santiago del Estero, Argentina

Martín Vázquez Centro Austral de Investigaciones Científicas (CADIC), CONICET, Ushuaia, Argentina

Lucía Watson Jiménez Universidad Nacional Autónoma de México, Mexico City, Mexico

Editors

Mariana Mondini Laboratorio de Zooarqueología y Tafonomía de Zonas Áridas (LaZTA), IDACOR, CONICET/Universidad Nacional de Córdoba, Córdoba, Argentina

Facultad de Filosofía y Letras, Universidad de Buenos Aires, Ciudad Autónoma de Buenos Aires, Argentina

A. Sebastián Muñoz Laboratorio de Zooarqueología y Tafonomía de Zonas Áridas (LaZTA), IDACOR, CONICET/Universidad Nacional de Córdoba, Córdoba, Argentina

Pablo M. Fernández Instituto Nacional de Antropología y Pensamiento Latinoamericano—CONICET, Ciudad Autónoma de Buenos Aires, Argentina

Facultad de Filosofía y Letras, Universidad de Buenos Aires, Ciudad Autónoma de Buenos Aires, Argentina

Peer-Reviewers

Analía Andrade Centro Nacional Patagónico (CENPAT), CONICET, Puerto Madryn, Argentina

Frank E. Bayham Department of Anthropology, California State University, Chico, CA, USA

Luis Borrero Instituto Multidisciplinario de Historia y Ciencias Humanas (IMHICIHU), CONICET, Buenos Aires, Argentina

Nicolas Goepfert Archéologies des Amériques (ARCHAM), CNRS, Paris, France

Sandra Gordillo Centro de Investigaciones en Ciencias de la Tierra (CICTERRA), CONICET/UNC, Córdoba, Argentina

Lisa Hodgetts Department of Anthropology, Social Science Centre, Western University, London, ON, Canada

Jean L. Hudson Department of Anthropology, University of Wisconsin-Milwaukee—Milwaukee Public Museum, Milwaukee, WI, USA

G. Lorena L'Heureux Instituto Multidisciplinario de Historia y Ciencias Humanas (IMHICIHU), CONICET, Buenos Aires, Argentina

Patricio López Mendoza ARQMAR—Depto. de Antropología, Facultad de Ciencias Sociales, Universidad de Chile, Santiago, Chile

Daniel Loponte CONICET—Instituto Nacional de Antropología y Pensamiento Latinoamericano (INAPL), Buenos Aires, Argentina

Lembi Lõugas Department of Archaeobiology and Ancient Technology, Institute of History, Archaeology and Art History, Tallinn University, Tallinn, Estonia

Agustina Massigoge Investigaciones Arqueológicas y Paleontológicas del Cuaternario Pampeano (INCUAPA), CONICET/UNICEN, Olavarría, Argentina

Matías E. Medina División Arqueología, Facultad de Ciencias Naturales y Museo, Universidad Nacional de La Plata—CONICET, La Plata, Argentina

Pablo Messineo Investigaciones Arqueológicas y Paleontológicas del Cuaternario Pampeano (INCUAPA), CONICET/UNICEN, Olavarría, Argentina

Katherine M. Moore Department of Anthropology, University of Pennsylvania, Philadelphia, PA, USA

Marta Moreno García Centro de Ciencias Humanas y Sociales (CCHS), CSIC, Madrid, Spain

David C. Orton BioArCh, Department of Archaeology, University of York, York, UK

Diego D. Rindel CONICET—Instituto Nacional de Antropología y Pensamiento Latinoamericano (INAPL), Buenos Aires, Argentina

Federico L. Scartascini Instituto Multidisciplinario de Historia y Ciencias Humanas (IMHICIHU), CONICET, Buenos Aires, Argentina

Geoff Smith MONREPOS—Archaeological Research Centre and Museum for Human Behavioural Evolution, Schloss Monrepos, Neuwied, Germany

Peter Stahl Department of Anthropology, University of Victoria, Victoria, BC, Canada

Mikel Zubimendi División Arqueología, Facultad de Ciencias Naturales y Museo, Universidad Nacional de La Plata—CONICET, La Plata, Argentina

Zooarchaeology in the Neotropics: An Introduction

Mariana Mondini, A. Sebastián Muñoz, and Pablo M. Fernández

This book brings together a collection of works on the archaeology of human-animal interactions through time in the Neotropical Biogeographic Region. This huge area, ranging from Central Mexico to Southern Patagonia, is characterized by an outstandingly rich biodiversity distributed across an amazing array of contrasting environments. Understanding the zooarchaeological imprint of human insertion in the rich and singular Americas is, thus, an opportunity for improving our knowledge of the many ways modern humans have dealt with the global colonization of our planet and of the diversity of subsequent organization forms within such diverse settings.

The Neotropical zoogeographic region was first recognized and defined by Sclater (1858) and Wallace (1876). Since then, it has been successively divided into distinct subregions, basically comprising the Caribbean islands, a highly diverse subregion(s) to the north and east, and an arid one(s) to the west and

M. Mondini (✉)
Laboratorio de Zooarqueología y Tafonomía de Zonas Áridas (LaZTA), IDACOR, CONICET/ Universidad Nacional de Córdoba, Av. H. Yrigoyen 174, 5000 Córdoba, Argentina

Facultad de Filosofía y Letras, Universidad de Buenos Aires, Ciudad Autónoma de Buenos Aires, Argentina
e-mail: mmondini@conicet.gov.ar

A.S. Muñoz (✉)
Laboratorio de Zooarqueología y Tafonomía de Zonas Áridas (LaZTA), IDACOR, CONICET/ Universidad Nacional de Córdoba, Av. H. Yrigoyen 174, 5000 Córdoba, Argentina
e-mail: smunoz@conicet.gov.ar

P.M. Fernández (✉)
Instituto Nacional de Antropología y Pensamiento Latinoamericano—CONICET, Ciudad Autónoma de Buenos Aires, Argentina

Facultad de Filosofía y Letras, Universidad de Buenos Aires, Ciudad Autónoma de Buenos Aires, Argentina
e-mail: pablomarcelofernand@gmail.com

south, as well as different transition zones (Hershkovitz 1958; Rapoport 1968; Cabrera and Willink 1980; Simpson 1980; Patterson and Timm 1987; among others; for historical reviews and recent proposals, see Cox 2001; Morrone 2001, 2014; Solari et al. 2012; Holt et al. 2013).

The outstanding variety of Neotropical environments and landscapes—ranging from extreme deserts to savannas and grasslands, from alpine tundra to tropical rainforests—encompasses a great range of biomes and potential niches, greater than those in the northern regions (MacDonald 2003; Patterson and Costa 2012). The abiotic properties of the region—including its geometry, physical configuration, latitude and oceanity—also impinge upon the particular configuration of its biota (Morello 1984).

The wide array of Neotropical faunas and their high levels of endemism relate to this diversity and to the geological history of the South American subcontinent (Redford and Eisenberg 1989, 1992, 1999; MacDonald 2003; Patterson and Costa 2012; and references therein). It has been an island continent for most of the last 65 million years, although intermittent contact with other continents produced biotic exchanges at different times, contributing to its past and present diversity. More recently, some 3 million years ago, the Panama isthmus was formed and prompted the Great American Biotic Interchange. This not only allowed the introduction of species from the north, but also led to the extinction of many marsupial species in the Neotropical region. Today, while bats prevail among mammals to the north, rodents do to the south, and terrestrial carnivores and marine mammals become more important in the Southern Cone (Redford and Eisenberg 1992). More than 1500 mammalian species live in the Neotropics at present, which comprise about 30% of all extant species in the globe (Patterson and Costa 2012).

Such environmental scenario and faunal diversity have been critical in shaping human insertion into the faunal community as they colonized the last landmass on Earth—besides Antarctica—and also in shaping the evolution of human-animal interactions ever since (Pineau et al. 2003; Muñoz and Mondini 2008a, b; Borrero 2008; Fernández et al. 2014). The particular configuration of these settings has prompted unique relationships with these diverse animals and has involved specific taphonomic processes. Different kinds of interaction, from competition to commensalism, developed between humans and animals. Some of the most intense relationships produced domesticated species, as is the case of some birds (the Muscovy duck, and the turkey in the Mexican transition zone), a rodent (the guinea pig) and two camelids (the llama and the alpaca) (Stahl 2008). In this volume, some instances of these varied human-animal interactions and of the processes forming the zooarchaeological record in the Neotropics are outlined, as are some ways to address their study.

The chapters in the following pages derive from some of the contributions presented at the Second Academic Meeting of the Neotropical Zooarchaeology Working Group of the International Council for Archaeozoology (NZWG-ICAZ), which took place at the 12th ICAZ International Conference held in San Rafael, Argentina, in September 2014. The meeting centered on exploring the particularities displayed by the Neotropical zooarchaeological record and on discussing the processes originating it and their consequences in the evolution and diversity of human-animal interactions from a global perspective. The mission of the NZWG-ICAZ is precisely to offer a forum where these research problems can be discussed and shared (see http://alexandriaarchive.org/icaz/workneotropical).

The topics covered in this volume shed light on different and complementary aspects of state-of-the-art zooarchaeological research into the Neotropics. Several chapters focus on marine resources, and this partly relates to the fact that a large part of the region is a peninsula within an oceanic hemisphere (Morello 1984). These chapters cover a broad range of the variation found in the Neotropical coastal environments. Martinoli and Vázquez deal with pinniped exploitation by hunting and gathering populations in temperate insular settings (Tierra del Fuego island) in Middle and Late Holocene contexts. They found contrasting ways of using *Arctocephalus australis* and, hence, contribute to the current understanding of human attitudes towards these marine mammals by broadening the range of variation known. Frontini and Bayon discuss the use of marine and coastal resources in different locations of the nearby southern Pampas (Buenos Aires province) in a similar time period. After reviewing resource representation in samples from coastal and inland settings, they discuss the use of this kind of resources through time and propose a differential use of marine items during the Holocene. Further to the north, Silva and colleagues provide a thorough account of Mid- to Late Holocene shell mounds from the southeastern coast of Brazil. The emphasis is not just on human behaviour but rather on shell mounds as proxies for biodiversity. In bringing together a wealth of malacological information, biodiversity patterns are inferred and discussed for the region.

Inland Neotropical faunas also have unique characteristics given the variety of environments they inhabit and the long history of isolation of the South American subcontinent. Another set of chapters deals primarily with these faunas—both terrestrial and riverine/estuarine, including birds—and also with varying societal organizations. Such is the case of the chapter by Escosteguy and Salemme, who study faunal exploitation by hunter-gatherers in Cañada Honda, a riverine setting in the Pampas, contributing to our knowledge of the diversity of human-animal interactions in the region. Increasing dietary diversification and intensification of small vertebrate exploitation during the Late Holocene is inferred. Beovide and colleagues discuss resource exploitation as recorded in shell middens found in an estuarine environment over the Río de la Plata during Mid-Late Holocene. They analyze spatial and temporal resource catchment, as well as the consequences of the introduction of pottery by 3000 years BP. On the other hand, Cardoza and colleagues account for a case study in the Pacific basin of Peru during the Mid-Archaic and the Initial Period. Pernil Alto, in the Palpa Valleys, is one of the few settlements known so far that is informative of human-animal interactions in the area and of this age. In the latter period, when sedentism was being established, increased emphasis on camelids is inferred.

Natural formation processes in Neotropical environments are also dealt with in this collection of works. Muñoz makes a taphonomic analysis of Late Holocene surface bone assemblages from southern Patagonia. He discusses natural bone modifications which could be informative of the transition between burial and exposure conditions in this kind of assemblages, which are abundant in the Atlantic face of coastal Patagonia. Also from a taphonomic perspective, the fossorial faunal record from the dry Chaco region in Santiago del Estero province, Argentina, is discussed by del Papa and colleagues. They differentiate individuals died of natural causes inside their burrows from those deposited by natural predators and in anthropic accumulations. Hence, a more precise interpretation of the role of burrowing rodents in human diets between 1200 AD and the Spanish conquest is offered.

Finally, Neotropical faunas also entail unique methodological challenges, and some chapters contribute new information from this perspective. Buckley and colleagues as well as Mondini and Muñoz deal with the taxonomic identification of Neotropical faunas; the former through collagen fingerprinting and the latter through osteometry. Buckley and colleagues explore the application of collagen fingerprinting analyses to remains of a dwarf deer of uncertain ancestry discovered in a ~6000 year-old shell-bearing midden in Pedro González Island (Panama), and discuss the taxonomic affinity of this and other deer in Central America and Amazonia. On the other hand, Mondini and Muñoz focus on the osteometrics of *Vicugna vicugna* and *Lama guanicoe* individuals from poorly known areas of their present range, and discuss variation recognition of Neotropical wild camelids.

Several other contributions were presented at the Second Academic Meeting of the NZWG-ICAZ apart from those included in this volume. The complete list of presentations can be found in the conference proceedings (see ICAZ 2014). At the meeting, we were honored with the discussion of the oral presentations by Dr. Susan deFrance (Department of Anthropology, University of Florida), who highlighted several aspects of these works, including interdisciplinarity and the application of sophisticated methods, as well as the need to link specific case studies to broader anthropological questions.

As a concluding remark, we would like to highlight that the chapters in this volume, along with the other presentations that contributed to the Second Academic Meeting of the NZWG-ICAZ, represent some instances of the variation in human-animal interactions through time in the Neotropics. They help grasp how unique they have been, and yet how much can be learnt from them even for other settings and other times. From a longer-term perspective, they reveal how much Neotropical zooarchaeology has been growing in the past few decades. It is our hope that it will continue to grow and become even stronger and, in so doing, it will most certainly reveal a varied array of interactions of all kinds with some of the most diverse faunas on Earth.

Acknowledgements We are sincerely grateful to all authors and reviewers. We are also thankful to all the participants in the Neotropical zooarchaeology session, to the session discussant, and to the 12th ICAZ International Conference organizers. Carolina Mosconi kindly revised the English of this introduction.

References

Borrero LA (2008) The archaeology of the Neotropics. Quat Int 180:152–157. In: Muñoz AS, Mondini M (eds) Neotropical zooarchaeology and taphonomy
Cabrera AL, Willink A (1980) Biogeografía de América Latina, Serie Monográfica13, Serie de Biología. Organización de Estados Americanos, Washington
Cox CB (2001) The biogeographic regions reconsidered. J Biogeogr 28:511–523
Fernández PM, Mondini M, Muñoz AS, Cartajena I (eds) (2014) Hacia una zooarqueología de los Neotrópicos. Etnobiol 12(2):1–94
Hershkovitz P (1958) A geographical classification of Neotropical mammals. Field Zool 36:581–646
Holt BG, Lessard JP, Borregaard MK, Fritz SA, Araújo MB, Dimitrov D, Fabre PH, Graham CH, Graves GR, Jønsson KA, Nogués-Bravo D, Wang Z, Whittaker RJ, Fjeldså J, Rahbek C (2013) An update of Wallace's zoogeographic regions of the world. Science 339:74–78
ICAZ (2014) Abstracts/Libro de Resúmenes, 12da Conferencia Internacional ICAZ/ICAZ 12th International Conference (San Rafael, Mendoza, Argentina, Septiembre/September 22nd-27th, 2014). Facultad de Filosofía y Humanidades, Universidad Nacional de Córdoba, Córdoba
MacDonald G (2003) Biogeography: introduction to space, time, and life. Willey, New York
Morello J (1984) Perfil Ecológico de Sudamérica. Características estructurales de Sudamérica y su relación con espacios semejantes del planeta. ICI-Ediciones Cultura Hispánica, Barcelona
Morrone JJ (2001) Biogeografía de América Latina y el Caribe. M & T Manuales y Tesis SEA, vol 3. CYTED, ORCYT-UNESCO and SEA, Zaragoza
Morrone JJ (2014) Biogeographical regionalisation of the Neotropical region. Zootaxa 3782 (1):1–110
Muñoz S, Mondini M (2008a) Long term human/animal interactions and their implications for hunter-gatherer archaeology in South America. In: Papagianni D, Layton R, Maschner HDG (eds) Time and change: archaeological and anthropological perspectives on the long term. Oxbow Books, Oxford, pp 55–71
Muñoz S, Mondini M (eds) (2008b) Neotropical zooarchaeology and taphonomy. Quat Int 180 (1):1–157
Patterson BD, Costa LP (2012) Introduction to the history and geography of Neotropical mammals. In: Patterson BD, Costa LP (eds) Bones, clones, and biomes: the history and geography of recent Neotropical mammals. University of Chicago Press, Chicago, pp 1–5
Patterson BD, Timm RM (1987) Studies in Neotropical mammalogy: essays in honor of Philip Hershkovitz, Field Zool, n.s, vol 39. Field Museum of Natural History, Chicago
Pineau V, Zangrando A, Scheinsohn V, Mondini M, Fernández P, Barberena R, Cruz I, Cardillo M, Muscio H, Muñoz AS, Acosta A (2003) Las particularidades de Sudamérica y sus implicaciones para el proceso de dispersión de *Homo sapiens sapiens*. In: Curtoni R, Endere ML (eds) Análisis, Interpretación y Gestión en la arqueología de Sudamérica, Serie Teórica No 2. INCUAPA, Facultad de Ciencias Sociales Universidad Nacional del Centro de la Provincia de Buenos Aires, Olavarría, pp 121–133
Rapoport EH (1968) Algunos problemas biogeográficos del Nuevo Mundo con especial referencia a la región Neotropical. In: Delamare-Debouteville CL, Rapoport EH (eds) Biologie de L'Amerique Australe, vol 4. Centre National du Recherche Scientifique, Paris, pp 55–110
Redford KH, Eisenberg JF (1989) Mammals of the Neotropics, Volume 1: The northern Neotropics: Panama, Colombia, Venezuela, Guayana, Suriname, French Guiana. University Chicago Press, Chicago

Redford KH, Eisenberg JF (1992) Mammals of the Neotropics, Volume 2: The Southern Cone: Chile, Argentina, Uruguay and Paraguay. University Chicago Press, Chicago

Redford KH, Eisenberg JF (1999) Mammals of the Neotropics, Volume 3: The Central Neotropics: Ecuador, Peru, Bolivia, Brazil. University Chicago Press, Chicago

Sclater PL (1858) On the general geographic distribution of the members of the Class Aves. J Linn Soc: Zool 2:130–145

Simpson GG (1980) Splendid isolation: the curious history of South American mammals. Yale University Press, New Haven

Solari S, Velazco PM, Patterson BD (2012) Hierarchical organization of Neotropical mammal diversity and its historical basis. In: Patterson BD, Costa LP (eds) Bones, clones, and biomes: the history and geography of recent Neotropical mammals. University of Chicago Press, Chicago, pp 145–156

Stahl PW (2008) Animal domestication in South America, The handbook of South American archaeology. Springer, New York, pp 121–130

Wallace AR (1876) The geographical distribution of animals. Macmillan, London

Pinniped Capture and Processing: A Comparative Analysis from Beagle Channel (Tierra del Fuego, Argentina)

María Paz Martinoli and Martín Vázquez

2.1 Introduction

The Beagle Channel is located on the southern coast of the Isla Grande de Tierra del Fuego (Fig. 2.1). It was inhabited by maritime hunter-gatherer-fishers from 6400 radiocarbon years BP to the late nineteenth century AD, when the European permanent settlement in the island began. Archaeological data have shown that these human groups had a diversified subsistence focused on marine resources, where pinnipeds provided the greatest amount of calories to the diet (Schiavini 1990, 1993; Orquera and Piana 1999, 2009; Orquera 2005; Zangrando 2003, 2009a, b; Tivoli and Zangrando 2011). However, recent zooarchaeological studies have revealed variations in the exploitation of resources among these prehistoric people during the Late Holocene: marine and terrestrial mammals decreased in order of importance in later assemblages, fish and bird remains increased in general faunal representation during the last 1500 years (Zangrando 2009a, b; Tivoli 2010a, b; Tivoli and Zangrando 2011).

While pinnipeds sex/age profiles and anatomical representation have been studied for the Middle Holocene (Schiavini 1990, 1993; Orquera and Piana 1999), we did not have such data from other archaeological contexts. Most of capture, processing and butchery patterns were not comprehensively analyzed in regional and supra-regional scale (Muñoz 2011). Moreover, the link between the long term changes of diet and exploitation modes of pinnipeds in the Beagle Channel remained unknown.

The aim of this study is therefore to evaluate exploitation strategies of pinnipeds excavated from shell middens at two different archaeological localities of the Beagle Channel with different ages: *Imiwaia I* (Middle Holocene) (Orquera and Piana 1999, 2000; Zangrando 2009a; Tivoli 2010a) and *Ajej I* (Late Holocene)

M.P. Martinoli (✉) • M. Vázquez
CADIC-CONICET, Bernardo Houssay 200, 9410 Ushuaia, Tierra del Fuego, Argentina
e-mail: mpmartinoli@yahoo.com.ar; vazquezmartin68@gmail.com

© Springer International Publishing AG 2017
M. Mondini et al. (eds.), *Zooarchaeology in the Neotropics*,
DOI 10.1007/978-3-319-57328-1_2

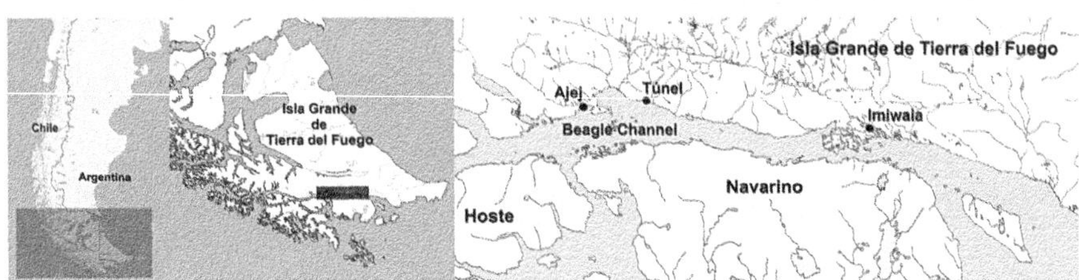

Fig. 2.1 Isla Grande de Tierra del Fuego, Beagle Channel. *Ajej*, *Túnel* and *Imiwaia* archaeological localities

(Piana et al. 2008). Pinniped capture and processing strategies were previously analyzed at site level in both locations, but a temporal evaluation of these activities is still needed.

Prey exploitation involves a set of interconnected activities between the time of carcass procurement and final disposal, and there are several variables that influence butchering decisions (Lyman 1992). When dealing with questions on pinnipeds hunting strategies, two factors are recognized as particularly important (Binford 1978; Lyman 1992, 2008; Hildebrandt and Jones 1992; Gifford-Gonzalez and Sunseri 2009) and they interact with each other: (a) foraging areas: pinnipeds occupy both the marine and terrestrial environments; and (b) prey sizes: pinnipeds have a distinct sexual dimorphism, which affect mainly animal size. In the case of the capture of large-sized prey on land, seals underwent primary disarticulation on kill sites and those parts considered of marginal value were probably abandoned at the hunting location (Binford 1978; Gifford-Gonzalez and Sunseri 2009). But when the capture took place in the water, combined with transport technology (canoes) and a residential area located nearby the foraging places, it would imply the complete carcass transportation back to the residential site regardless of prey size (Lyman 1992; Orquera and Piana 1999; Ames 2002). On the contrary, small prey tends to be transported complete to the final consumption place, in spite of its distance from the foraging areas (Binford 1978). Based on the previous assumptions we can generate a particular set of archaeological expectations regarding three interrelated aspects: sex and age profiles, anatomical representations and butchery marks on bones.

2.1.1 Pinnipeds as Resource for Hunter-Gatherer-Fishers in the Beagle Channel

The main pinniped species in Tierra del Fuego are the South American fur seal (*Arctocephalus australis*) and the Southern sea lion (*Otaria flavescens*) (Bastida and Rodríguez 2003). The former is the most abundant in the archaeological record of the north coast of the Beagle Channel. The weight recorded for South American fur seals ranges between 150 and 200 kg for males and 60 kg for females (King 1983), but the weight recorded in Uruguay's population spans between 80 and

60 kg for males and with an average of 40.6 kg for females (Schiavini 1990, 1993). These otariids have a polygamous behaviour and their annual cycle is divided between a short reproductive stage and a period of regular visits to coastal areas (Crespo et al. 2008).

The age of sexual maturity is about 3 years for female seals and 7 or 8 years for male seals. The mating season of *A. australis* is during summer (Sielfeld 1983, 1999), and they are available in breeding colonies. Throughout the rest of the year, adult and young male seals spend more time foraging in the sea whereas adult female seals must return regularly to the colonies to nurse their pups (King 1983; Campagna 1985). South American fur seal colonies are located on rocky shores (Sielfeld 1983), and most of the breeding colonies are located in outer coasts and islands of the archipelago (Schiavini 1990; Crespo et al. 2008, p. 2), such as *Isla de los Estados*, *Isla Observatorio*, or in the surroundings of *Cape Horn* (Schiavini and Raya Rey 2001). However, it has been documented haul-outs in the Beagle Channel (Schiavini and Raya Rey 2001; Crespo et al. 2008, p. 3).

Regarding pinniped exploitation during the Holocene, the analysis of the layer D of the *Túnel I* (6400–4500 BP) site provided most of the archaeological data to build a general model of pinnipeds capture and processing (Schiavini 1990, 1993; Orquera and Piana 1999). The NISP (Number of Identified Specimens) value for pinniped remains from this layer is 59,300 (65% of the D layer total NISP) (Orquera 2015, pers. comm.). Schiavini (1990, Table 29) has determined MNI (Minimum Number of Individuals) values of 273 for *A. australis* and 9 for *O. flavescens* considering maxillae and mandibles. This author has also identified sex and age, based on the canines, for 223 individuals of *A. australis*: 86.5% of them were males and 69% of these males were under 8 years of age (non-reproductive males), whereas in females 37% of the bones came from animals of less than 4 years of age (Schiavini 1990). All anatomical units of these individuals are represented, although the frequencies have not been published (Orquera and Piana 1999). According to the study of maxillary canines it was determined that most of *A. australis* represented in the layer D (90%) died between March and September (Schiavini 1990, Fig. 42).

These zooarchaeological studies based on determination of sex, age, season of death and anatomical profiles, led to two main interpretations. First, the zooarchaeological assemblages of layer D of *Túnel I* are composed mainly by males (83%) of *A. australis* killed between autumn and spring (Schiavini 1990, 1993). Considering that the rockeries are mainly located in outer parts of the archipelago, it was proposed that pinniped captures should have occurred predominantly in the water, foraging in the sea with canoes and harpoons. Second, it was proposed that entire carcasses were transported and butchered in consumption places (Orquera and Piana 1999).

The Second Component of *Túnel I* has a Middle Holocene archaeological context. As said above, in the excavated sites younger than 1500 years BP the relative importance of pinnipeds in the diet has decreased (although they continued to provide the largest amount of calories to the human diet), while the importance of offshore preys (fish and birds) has increased (Zangrando 2009a, b; Tivoli and Zangrando 2011). So, one may ask if such a change in subsistence strategy correlated also with changes in the modes of exploitation of pinnipeds. The two proposed possible changes are: differences in foraging areas (foraging in the water/ foraging near the colonies) and consequently differences in prey choice (males/ females). These possible differences generate in turn these specific expectations:

(a) Changes in the foraging areas imply differences in prey selection: pinnipeds can be considered as highly predictable preys, both spatially and temporally (Lanata and Borrero 1994) and they have ruled behaviours according to season, sex and age. Thus, sex and age profiles represented in the bone assemblages can be informative about the possible hunter-gatherer foraging areas (Lyman 1989, 2003): the capture of isolated individuals in the water should result in a profile where adult and subadult males dominate, while the predominance of females sexually matured and pups would indicate the exploitation of colonies.

(b) Changes in prey selection probably imply different decisions regarding transport and carcass butchering: pinnipeds have a distinct sexual dimorphism; males double the size of females. According to traditional understanding one of the main reasons of differential transportation is prey size (Binford 1978; Hawkes and O'Connell 1985). Therefore we assume a transport of adult females and subadults/juveniles in complete carcasses while adult males would have been transported in incomplete carcasses at *Imiwaia I* and *Ajej I*. But another two aspects in transport decisions are: transport distance (Binford 1978) and transport technology (Ames 2002). Both sites were located in coastal areas, so if some of the preys were actually captured in the water using canoes we would assume a low transportation cost (Ames 2002), and therefore complete anatomical profiles. But if the animals were captured within the colonies we would expect different transport strategies depending on prey size (Gifford-Gonzalez and Sunseri 2009). We would also expect differences in the nature and amount of butchery marks depending on the implied distance in carcasses transport (Binford 1978).

2.2 Materials

2.2.1 Imiwaia I

The *Imiwaia I* site (layers M, L and K; 6000–4500 years BP) is located in the *Cambaceres* Bay (54°52′26″S, 67°17′59″W). It is a multicomponent site with a central depression (house pit) surrounded by shell midden deposits (Orquera and Piana 2000). This site was interpreted as a residential locus, where multiple activities took place (Orquera and Piana 1999, 2000).

Nearly 36,000 bone specimens were recovered from layers M, L and K; most of which were identified taxonomically (NISP = 32,424). Figure 2.2 shows a clear predominance of fish (NISP = 20,367; 63%), followed by birds (NISP = 5343;

Fig. 2.2 Faunal representation in layer M, L and K (Martinoli 2015)

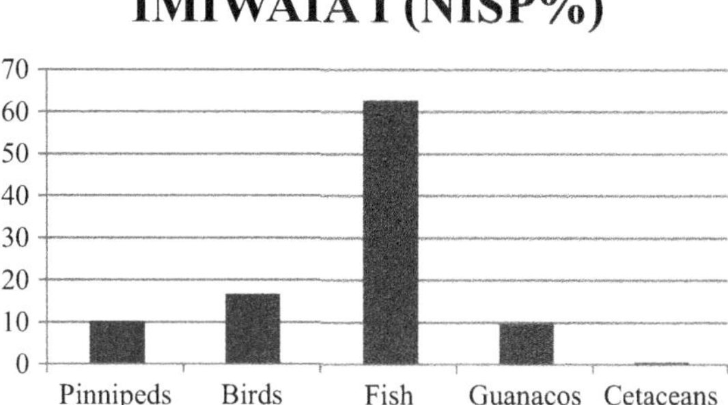

Fig. 2.3 Faunal representation in layer C (Piana et al. 2008)

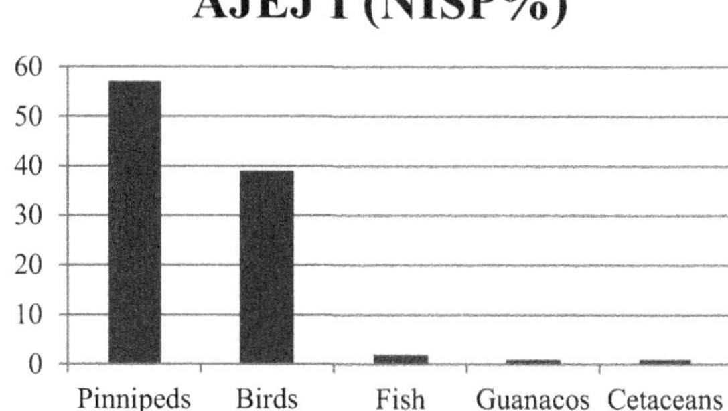

16%). Pinnipeds come in the third order of importance (NISP = 3316; 10%). Guanacos and cetaceans have the lowest representation (Martinoli 2015).

2.2.2   *Ajej I*

The *Ajej I* site (layer C; 1400 years BP) was an isolated shell midden, currently located at 30 m of *Pipo* River (54°50′19.5″S, 68°21′37.4″W). The study of this site was a result of rescue excavations carried out in the northern coast of the Beagle Channel during March 1999.

The external appearance of *Ajej I* was a small mound which reached 30 cm above the surrounding surface. *Ajej I* was a monocomponent shell midden with a low redundant occupation where restricted activities were identified (Piana et al. 2008).

A total of 2106 animal bones were recovered from the shell midden (layer C); over 80% were identified taxonomically (NISP = 1808). Figure 2.3 shows a clear predominance of pinnipeds (NISP = 1011; 57%), followed by birds (NISP = 499; 39%). Remains of fish, guanaco and cetaceans have very low representations (Piana et al. 2008).

One finding of a pinniped thoracic vertebra (juvenile *A. australis*) is noteworthy since it has a lithic projectile point incrusted in its vertebral body (Piana et al. 2008, Fig. 2.7). This lithic point was interpreted as an arrowhead by its morphology, weight and size; also, it is similar to some arrowheads described in ethnographic sources (Piana et al. 2008).

2.3 Results

The zooarchaeological analyses indicate that *A. australis* is the most abundant otariid species in both zooarchaeological assemblages described above (Table 2.1). However, there is an important difference concerning the age and sex profiles of this animal in both sites. In the *Imiwaia I* site, according to the determinations of long bones, adult males (MNI = 14) and subadults predominate (MNI = 10) (Martinoli 2015), while the analysis of growth structures of maxillary canines in *Ajej I* display a prevalence of sexually mature females (MNI = 8, all of them between 3 and 7 years old) (Piana et al. 2008).

2.3.1 Anatomical Representation

In layers M, L and K of *Imiwaia I A. australis* is represented by all anatomical elements, though not in the same frequency (Table 2.2). The anatomical region, which presents the highest frequency measured by %MAU (Minimal Animal Units), is the axial skeleton: the lumbar vertebrae (92%), scapulae (82%), sternebrae (75%) and pelves (75%). These elements are followed by abundance of limbs, which also show a high representation. The forelimbs are present by 74% (humeri), 93% (radii) and 100% (ulnae), and the hindlimbs by 87% (femuri), 51% (tibiae) and 71% (fibulae). Anatomical elements with lower rates of representation are the metatarsals (45%), carpals (36%) and tarsals (19%) (Martinoli 2015).

In contrast, *Ajej I* material shows an incomplete skeletal profile of *A. australis* (Fig. 2.4). The representation of the axial skeleton is highly variable (Table 2.2). Indeed, while there are anatomical units represented by values close to 50% (57% sternebrae, 44% pelves and 40% atlases), there are also elements with percentages less than 25% (8% scapulae, 22% cervical vertebrae and 18% dorsal vertebrae) or even absent like the lumbar-sacral sector. The limbs were represented by femuri

Table 2.1 Species representation by sex and age profiles (MNI)

| | O. flavescens (MNI) | | | | A. australis (MNI) | | | | | |
| | Adult | | Subadult | | Adult | | Subadult | | | Neonate | |
	Male	Female	Male	Female	Male	Female	Male	Female	Indet.	Indet.	Total
Imiwaia I	1	2	1	0	14	3	3	1	10	1	36
Ajej I	0	0	0	0	1	8	1	1	0	1	12

Table 2.2 Data of pinniped archaeofaunal assemblages according to NISP and MNE (Minimal Number of Elements) representation and anatomical representation (MAU and %MAU) from the *Ajej I* and *Imiwaia I* sites (Piana et al. 2008; Martinoli 2015)

Anatomical units	*Ajej I* (layer C)				*Imiwaia I* (layers M, L and K)			
Head	NISP	MNE	MAU	%MAU	NISP	MNE	MAU	%MAU
Inferior maxillary	28	24	12.5	**100**	33	27	13.5	49.1
Superior maxillary	2	2	1	8	–	–	–	–
Teeth	120	120	3.7	29.6	169	169	5.3	19.3
Cranium	53	7	7	56	–	–	–	–
Total	203	153	–	–	202	196	–	–
Axial skeleton	**NISP**	**MNE**	**MAU**	**%MAU**	**NISP**	**MNE**	**MAU**	**%MAU**
Scapulae	2	2	1	8	157	45	22.5	81.8
Ribs	157	104	4.3	34.4	526	453	18.9	68.7
Sternebrae	51	50	7.1	56.8	144	144	20.6	74.9
Pelves	11	11	5.5	44	62	41	20.5	74.5
Cervical vertebrae	55	29	4.1	32.8	140	134	19.1	69.4
Dorsal vertebrae	47	27	2.2	17.6	72	70	5.8	21.1
Lumbar vertebrae	–	–	–	–	141	127	25.4	92.4
Indeterminate vertebrae	–	–	–	–	7	7	0.3	1.1
Baculum	–	–	–	–	7	7	7	25.4
Total	323	223	–	–	1256	1028	–	–
Forelimbs	**NISP**	**MNE**	**MAU**	**%MAU**	**NISP**	**MNE**	**MAU**	**%MAU**
Humeri	4	3	1.5	12	78	41	20.5	74.5
Radii	18	9	4.5	36	97	55	27.5	**100**
Ulnae	15	6	3	24	75	51	25.5	92.7
Phalanges I (forelimbs)	9	9	4.5	36	23	23	11.5	41.8
Phalanges (forelimbs)	54	44	3.4	27.2	373	289	10.3	37.4
Carpals	28	28	4.7	37.6	138	138	9.8	35.6
Metacarpals	48	39	7.8	62.4	220	202	20.2	73.4
Total	176	138	–	–	1004	799	–	–
Hindlimbs	**NISP**	**MNE**	**MAU**	**%MAU**	**NISP**	**MNE**	**MAU**	**%MAU**
Femuri	22	10	5	40	72	48	24	87.3
Tibiae	18	10	5	40	48	28	14	50.9
Fibulae	10	6	3	24	39	39	19.5	70.9
Tarsals	34	34	2.4	19.2	106	106	15.1	54.9
Patellae	8	8	4	32	15	14	7	25.4
Phalanges I (hindlimbs)	9	9	4.5	36	17	17	8.5	30.9

(continued)

Table 2.2 (continued)

Anatomical units	*Ajej I* (layer C)				*Imiwaia I* (layers M, L and K)			
Head	NISP	MNE	MAU	%MAU	NISP	MNE	MAU	%MAU
Phalanges (hindlimbs)	76	61	4.7	37.6	373	289	10.3	37.4
Metatarsals	48	43	8.6	68.8	170	170	17	61.8
Total	225	181	–	–	840	711	–	–
Indeterminate	84	–	–	–	–	–	–	–
Total	1011	696	–		3302	2734	–	–

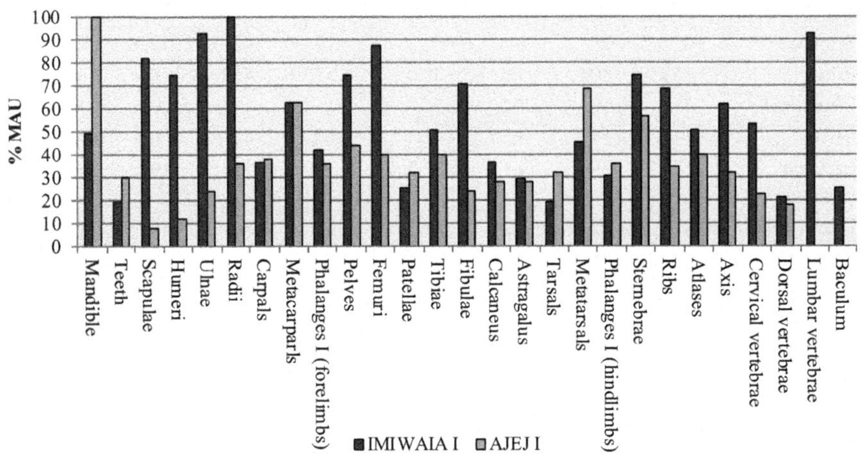

Fig. 2.4 Representation of anatomical units (%MAU)

(40%), tibiae (40%), radii (36%), ulnae (24%), fibulae (24%) and humeri (12%) (Piana et al. 2008).

In order to explore further cultural explanations for the different representation of anatomical parts, we have used a Meat Utility Index (%MUI) derived from an Otariid subadult male (San Román 2009). The assemblages %MAU values were correlated with otariid %MUI and the results indicate that in both cases the %MUI values are weakly correlated with the frequencies of pinnipeds remains (*Ajej I* $r_s = 0.19$ p > 0.05; *Imiwaia I* $r_s = -0.4$ p > 0.1), and are not statistically significant.

2.3.2 Butchery Marks

The data on bone surface modifications display some differences in processing activities between both assemblages. First, pinniped remains from *Imiwaia I* exhibit quantitatively less anthropogenic modifications than the bones from *Ajej I*. As it can be seen in Fig. 2.5, the forelimb elements with butchery marks formed less than 15% in *Imiwaia I* (6% humeri, 6% radii, 5% ulnae and 13% metacarpals). In contrast, the majority of forelimb elements from *Ajej I* have some kind of butchery marks (75% humeri, 50% radii, 20% ulnae and 44% metacarpals). The same relation is observed for the hindlimbs: while the former assemblage has a low frequency of marks (5% femuri, 1% tibiae, 1% fibulae and 7% metatarsals), the latter has a higher amount of bone modifications (36% femuri, 4% tibiae, 10% fibulae and 65% metatarsals). The axial skeleton is an exception, especially the sternebrae and ribs, because in both assemblages only few of these anatomical elements exhibit this type of marks.

Second, the frequency of butchery marks is not only higher in *Ajej I*, but also there are differences in the type of anthropogenic modifications present. While *Ajej I* assemblage exhibits cut marks, chop marks, scrape marks, and fresh bone fractures, most of the bones in *Imiwaia I* just have cut marks.

Regarding processing activities in *Imiwaia I*, tool marks mainly indicate the disarticulation of the limbs and axial skeleton. Most of the butchery marks located over the proximal and distal diaphysis of long bones, epiphysis (Fig. 2.6), and spinous processes of vertebrae were identified as due to dismemberment; although in low frequency the carcasses display the whole range of butchering activities. In previous work, it was suggested that the complete processing of the pinniped carcasses took place at the residential site, based on the anatomical profile added to identify butchering pattern (Martinoli 2015).

Bone specimens of the *Ajej I* site exhibit cut, scrape and chop marks over articular areas of limb bones and ventral sides of vertebrae. This pattern most likely responds to activities of carcass reduction into smaller portions and defleshing, and could also be due to differential transport of anatomical portions to the site (Piana et al. 2008).

However, one of the clearest differences between these two assemblages are the processing marks on flippers. Most of the metapodials and phalanges in the *Ajej I*

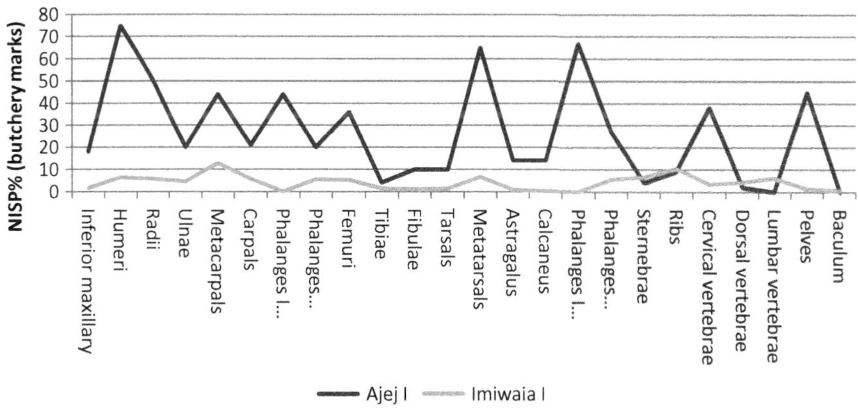

Fig. 2.5 Relative frequency of butchery marks on pinnipeds bones

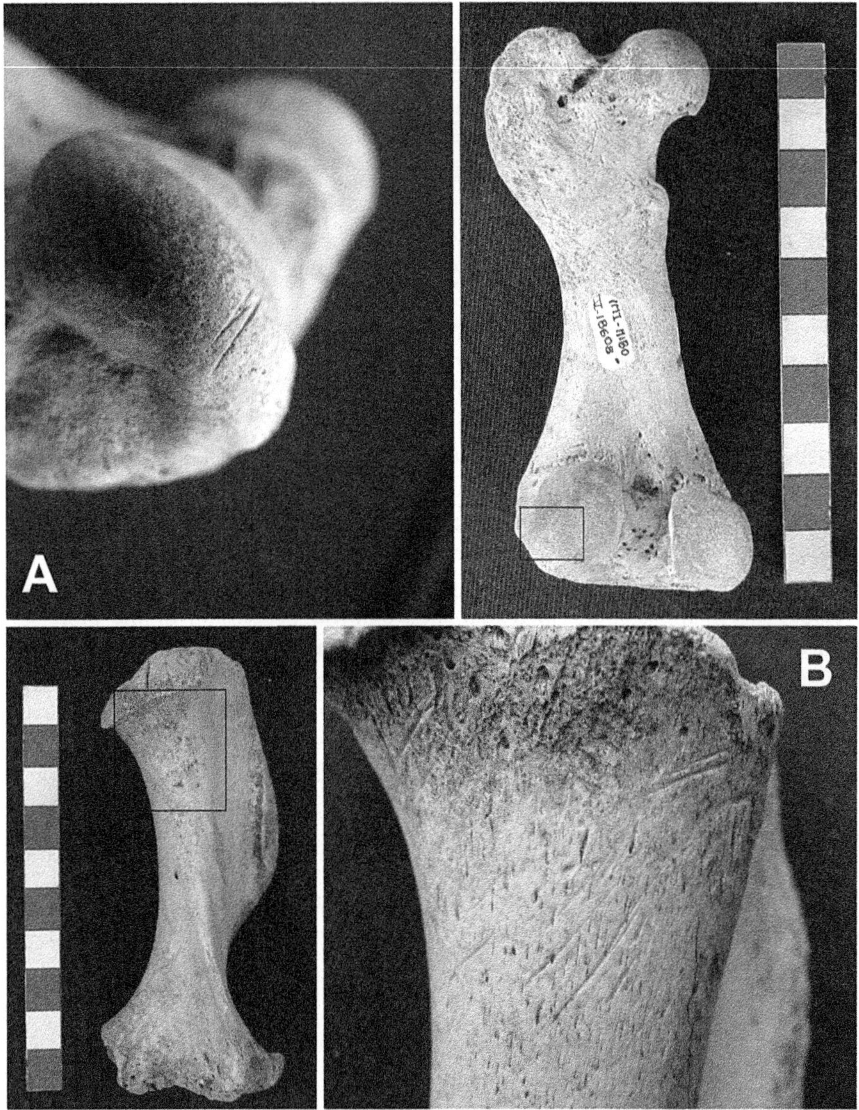

Fig. 2.6 *Imiwaia I*: cutmarks over distal epiphysis of femur (**a**) and proximal diaphysis of humerus (**b**)

site have butchery marks (54% metacarpals, 72% metatarsals, 33% phalanges), whereas the same anatomical elements in the *Imiwaia I* site display less anthropogenic modifications (25% metacarpals, 15% metatarsals and 7% phalanges). Furthermore, as can be seen in Fig. 2.7, the flipper bones of *Ajej I* site have chop marks (42%) and cut marks (58%), but *Imiwaia I* exhibits cut marks only. Prior

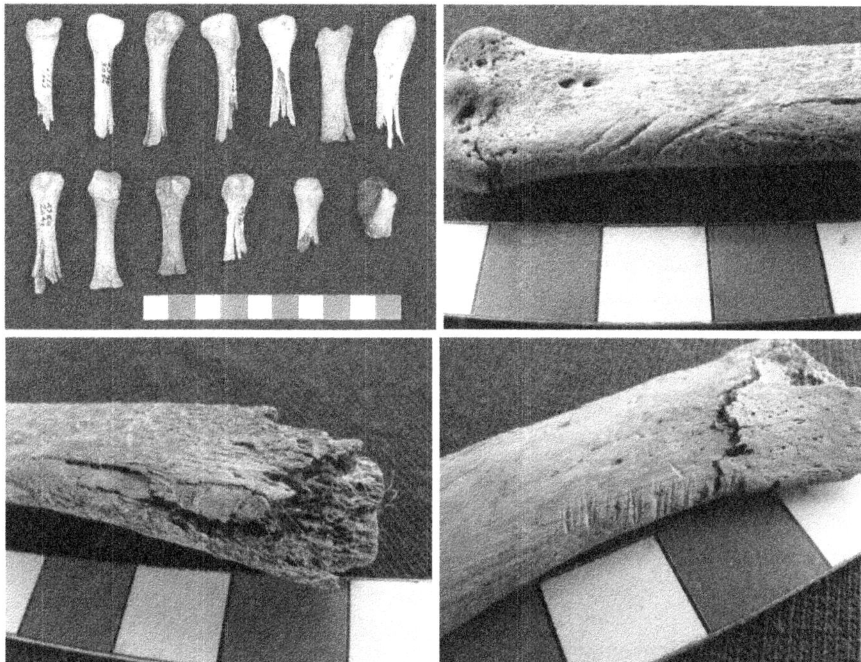

Fig. 2.7 *Ajej I*: cutmarks and chopmarks in phalanges, metacarpals and metatarsals

researchers have interpreted this pattern as a result of sectioning the distal region of the limbs to consume the fat accumulated in the flippers (Piana et al. 2008). However, although the zooarchaeological analyses of both sites show that the flippers were butchered for fat consumption, the *Ajej I* site has shown quantitative and qualitative differences in the processing activities.

Processing of radius, on the contrary, appeared to be very similar at both sites. These bones not only show cut marks, but also percussion marks and fresh bone fractures. As can been seen in Fig. 2.8, the radii were fractured in the medial and distal regions and the majority of the radii in both sites were affected. These elements also exhibited cut marks, and in some cases they were burned.

These similarities across assemblages might suggest that pinnipeds anatomy and the restrictions it imposes on carcass processing could have played a more important role in the butchery process of this portion than other kind of cultural constraints. This is due to the fact that in the body of pinnipeds some of the most important flipper muscles and tendons are in the radii medial diaphysis downwards (Cádegan Sepúlveda 2013, p. 54).

Regarding the burnt bones, the older assemblage shows a higher percentage (4%) than the younger one (1%). The difference was particularly in the sternum and the flippers: while the *Ajej I* site exhibited more than 25% of burnt specimens (Piana et al. 2008), the *Imiwaia I* site less than 2% (Martinoli 2015).

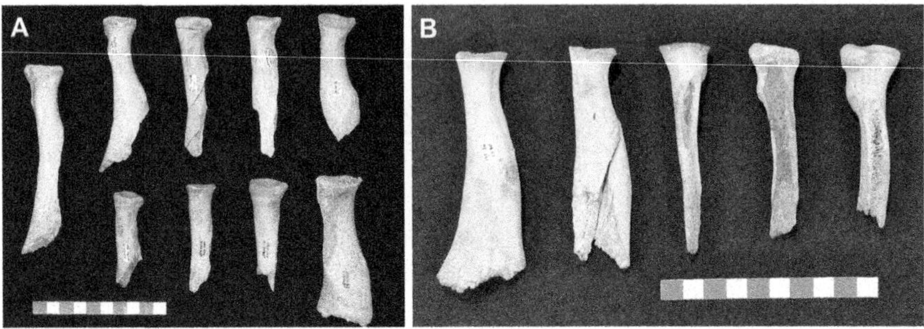

Fig. 2.8 Fresh bone fractures of radii from (**a**) *Imiwaia I* and (**b**) *Ajej I*

2.4 Discussion

In both assemblages the bone surfaces were very well preserved: in the *Imiwaia I* site most of the bones (92%) display 0 and 1 weathering stages (*sensu* Behrensmeyer 1978). Similarly, in the *Ajej I* site 80% of the pinniped bone remains exhibit 0 and 1 weathering stages (Piana et al. 2008). In both sites, these remains also present low fragmentation (Piana et al. 2008; Martinoli 2015). Thus, it cannot be possible to attribute the differences noticed above as a factor linked to differential preservation between assemblages.

Three important observations arise from the comparison between the *Imiwaia I* and *Ajej I* bone assemblages. The first aspect concerns the difference between the sex and age profiles. While in *Ajej I* most of the *A. australis* individuals are females in reproductive age, in *Imiwaia I* adult males and subadults of indeterminate sex of *A. australis* predominate. Given the known pinnipeds behaviour, the sex and age profiles of *Imiwaia I* assemblage—which are similar to the Second Component of *Túnel I*—supports the view that most of the identified preys were captured in the water, mainly isolated individuals foraging in the Beagle Channel. This kind of captures could have been carried out through the use of harpoons with detachable points and canoes (Schiavini 1990, 1993; Orquera and Piana 1999). On the contrary, in *Ajej I* there are a high proportion of females sexually mature. These animals do shorter trips than males to find food because they must return to the colonies to nurse their pups and therefore they spend most of the time on land.

The similarity between the mortality profiles from *Túnel I* and *Imiwaia I* (early assemblages) is remarkable, as well as the resemblance between the more recent assemblages of *Ajej I* and *Túnel VII*. In the latter site, there is also a predominance of pinniped subadults and females (Estévez et al. 1995; Zangrando 2009a; Zangrando et al. 2010). Hence, later sites show a different capture strategy: the high amount of females would imply that the hunter-gatherers were probably foraging near the colonies. Besides, *Ajej I* and *Túnel VII* (Estévez et al. 1995) have a pinniped thoracic vertebra with a lithic projectile point incrusted in the vertebral body, which implies also variability in capture strategies.

The second aspect regarding the differences between the sites refers to the distinct anatomical representations. In *Imiwaia I* most of the anatomical units are represented in similar proportions, much like in the Second Component of *Túnel I* (Orquera and Piana 1999). It was proposed then that the animals were carried entirely to the site, where they underwent primary disarticulation (Martinoli 2015). In contrast, anatomical profiles of the *Ajej I* material are incomplete, being the lumbar-sacral sector totally absent (Piana et al. 2008). If we assume that body size is one of the most important variables made in decisions about transportation it is remarkable that the site with smaller preys (*Ajej I, A. australis* females) exhibits a more incomplete anatomical profile, contrary to the archaeological expectations derived from the transport models (e.g. Binford 1978, 1981). Moreover, the correlation between %MAU and %MUI is weak and not statistically significant, meaning that the economic value of the anatomical portions did not influence the selection of body parts for transport to the consumption location. However, there are also other processes which may affect the body parts profile: namely social processes, i.e. cooperation mechanisms such as food sharing among households, where the social relationships mostly determine the sharing of body parts rather than its meat/ fat content (Winterhalder 1986; Marshall 1993).

Finally, the third aspect of differentiation between the *Ajej I* and *Imiwaia I* sites are the butchery activities. Employing the measurement of the co-occurrence between different variables such as the frequency of bone modifications, the morphology of the marks and its distribution over the bones and the carcasses (Fisher 1995; Egeland 2003; Pickering and Egeland 2006), it is important to note that the processing activity in *Ajej I* site was more intensive. Two of the clearest differences between these two sites are, on the one hand, the butchery of the flippers, which are portions with a low content of meat but a relative large amount of fat; and, on the other hand, the sternum, which has a high meat value in otariids (San Román 2009) and most of the sternebrae are burnt. Fresh bone fractures of the radii, similar at both sites, seem to be the consequence of pinnipeds anatomy.

Although it is necessary to incorporate the analysis of other bone assemblages of different age and context for a better understanding of the variability in human-pinnipeds relationships, the results obtained so far are relevant to highlight at least three aspects:

(a) *Variability in sex and age profiles*: the dominance of sexually mature females in *Ajej I* would indicate foraging near the colonies, which would probably imply a different capture strategy to that of *Imiwaia I* (e.g. the coordinated participation of a larger group of people; the capture of several preys into the same hunting event, among others).

(b) *Anatomical representation*: the site with smaller preys (females) display an incomplete body parts profile, but the selection of the parts is not related with their economical value, but perhaps with carcass location.
(c) *Butchery activities*: the intensity of carcass processing in *Ajej I* involves higher labour investment (processing costs) in butchery activities than in *Imiwaia I*.

On a regional scale, it was proposed that the use of landscape and resources varied through time in the Beagle Channel. The model suggests a *diversification* and *intensification* process: between 6000 and 1500 radiocarbon years BP, the archaeofaunal evidence shows a dominance of marine mammals over other resources. After the 1500 radiocarbon years BP, an increase in the representation of birds and fish was observed, involving an extension to offshore sectors for resource procurement (Zangrando 2009a, b; Tivoli 2010a, b; Tivoli and Zangrando 2011). The aspects previously outlined support some of the expectations of the model proposed by Zangrando (2009a, p. 108), where *"decreasing of the economic autonomy of households and increasing of the cooperativity in procurement activities"* (MP Martinoli translation) were expected under intensification conditions.

2.5 Final Remarks

In sum, the pinniped capture and processing patterns described in this paper display some important differences between the *Ajej I* and *Imiwaia I* sites. First, the different sex and age profiles imply variability in prey choice, and therefore differences in capture strategies. Second, while the later site presents an incomplete anatomical profile and a higher carcass processing intensity, in the earlier site most of the anatomical units are represented in similar proportions and exhibit low levels of carcass butchery, where the cutmarks mainly indicate primary disarticulation. This evidence agrees with the previously identified capture and transportation patterns from layer D of *Túnel I* for the Middle Holocene; however the exploitation of pinnipeds in the Late Holocene site displays differences in butchering activities and capture strategies.

Assemblages compared in this article have different chronologies, although it is not possible to establish that the reported differences are *exclusively* related with regional long term changes in the hunter-gatherer-fishers subsistence. However, the *Ajej I* site display a different pinniped capture and processing strategy than the one observed for the Second Component of *Túnel I*; whereas basal layers of *Imiwaia I*, with similar chronology to the latter site, exhibits a similar strategy to that of *Túnel I*. Still the range of variability in pinnipeds exploitation throughout the archaeological sequence in the Beagle Channel region remains unknown, but in this investigation we were able to recognize at least two different strategies.

Acknowledgements We would like to thank Francisco Zangrando and L. Orquera for reviewing earlier versions of this chapter. We also thank Sofia Tecce to help us with English corrections, the reviewers for their contributions, and especially we want to thank the editors for inviting us to write this article.

References

Ames K (2002) Going by boat. The forager-collector continuum at sea. In: Fitzhugh B, Habu J (eds) Beyond foraging and collecting: evolutionary change in hunter-gatherer settlement systems. Kluwer Academic/Plenum Publishers, New York, pp 19–52

Bastida R, Rodríguez D (2003) Mamíferos marinos de Patagonia y Antártida. Vázquez Manzini Editores, Buenos Aires

Behrensmeyer A (1978) Taphonomic and ecologic information from weathering. Paleobiology 4:150–162

Binford L (1978) Nunamiut ethnoarchaeology. Academic Press, New York

Binford L (1981) Bones: ancient men and modern myths. Academic Press, New York

Cádegan Sepúlveda K (2013) Anatomía comparada del esqueleto apendicular de dos especies de otáridos, *Otaria flavescens* (Shaw, 1800) *Arctophoca australis gracilis* (Zimmerman, 1783). Dissertation, Universidad Nacional de Chile

Campagna C (1985) The breeding cycle of the southern sea lion, *Otaria byronia*. Mar Mamm Sci 1 (3):210–218

Crespo E, García N, Dans S, Pedraza S (2008) *Arctocephalus australis*. In: Crespo E, García N, Dans S, Pedraza S (eds) Atlas de Sensibilidad Ambiental de la Costa y el Mar Argentino. Mamíferos Marinos, pp 1–9

Egeland C (2003) Carcass processing intensity and cutmark creation: an experimental approach. Plains Anthropol 48:39–51

Estévez J, Juan-Muns N, Martínez J, Piqué R, Schiavini A (1995) Zooarqueología y Antracología: estrategias de aprovisionamiento de los recursos animales y vegetales en Túnel VII. In: Estévez J, Vila A (eds) Treballs d'Etnoarqueologìa 1, Encuentros en los conchales fueguinos. CSIC/Universidad Autónoma de Barcelona, Barcelona, pp 143–238

Fisher J Jr (1995) Bone surface modifications in zooarchaeology. J Archaeol Method Theory 2 (1):7–68

Gifford-Gonzalez D, Sunseri C (2009) An earlier extirpation of fur seals in the Monterey Bay region: recent findings and social implications. Proc Soc Calif Archaeol 21:89–102

Hawkes K, O'Connell J (1985) Optimal foraging models and the case of the !Kung. Am Anthropol 87:401–404

Hildebrandt W, Jones T (1992) Evolution of marine mammal hunting: a view from the California and Oregon Coast. J Anthropol Archaeol 11(4):360–401

King J (1983) Seals of the world. In: British museum (natural history). Oxford University Press, Oxford

Lanata J, Borrero L (1994) Riesgo y Arqueología. Arqueología de Cazadores-Recolectores, límites, casos y apertura. Arqueol Contemp 5:129–142. Lanata J, Borrero L, editors

Lyman R (1989) Seal and sea-lion hunting: a zooarchaeological study from the Southern Northwest Coast of North America. J Anthropol Archaeol 8:68–99

Lyman R (1992) Prehistoric seal and sea-lion butchering on the Southern Northwest Coast. Am Antiq 57(2):246–261

Lyman R (2003) Pinniped behavior, foraging theory, and the depression of metapopulation and nondepression of a local population on the Southern Northwest Coast of North America. J Anthropol Archaeol 22(4):376–388

Lyman R (2008) Quantitative paleozoology. Cambridge University Press, Cambridge

Marshall F (1993) Food sharing and the faunal record. From bones to behavior: ethnoarchaeological and experimental contributions to the interpretation of faunal remains. Occas Pap 21:228–246

Martinoli MP (2015) Procesamiento y consumo de pinnípedos: el caso de las ocupaciones canoeras tempranas del sitio Imiwaia I (Tierra del Fuego, Argentina). Intersecciones en Antropología 16(2): 367–381

Muñoz A (2011) Pinniped zooarchaeological studies in Southern Patagonia: current issues and future research agenda. In: Bicho N, Haws J, Davis L (eds) Trekking the shore: changing coastlines and the antiquity of coastal settlement. Springer, New York, pp 305–331

Orquera L (2005) Mid-Holocene littoral adaptation at the southern end of South America. Quat Int 132:107–115

Orquera L, Piana E (1999) Arqueología de la región del canal Beagle (Tierra del Fuego, República Argentina). Soc Argent Antropol, Buenos Aires

Orquera L, Piana E (2000) Imiwaia 1: un sitio de canoeros del sexto milenio A.P. en la costa norte del canal de Beagle. In: Desde el País de los Gigantes. Perspectivas Arqueológicas de la Patagonia, vol II. Universidad de la Patagonia Austral, Río Gallegos, pp 441–453

Orquera L, Piana E (2009) Sea nomads of the Beagle Channel in southernmost South America: over six thousand years of coastal adaptation and stability. J Island Coast Archaeol 4:1–21

Piana E, Vázquez M, Álvarez M (2008) Nuevos resultados del estudio del sitio Ajej I: un aporte a la variabilidad de estrategias de los canoeros fueguinos. Runa 29:101–121

Pickering T, Egeland C (2006) Experimental patterns of hammerstone percussion damage on bones: implications for inferences of carcass processing by humans. J Archaeol Sci 33:459–469

San Román M (2009) Anatomía económica de *Otaria flavescens*. In López P, Cartajena I, García C, F Mena (eds) Zooarqueología en el confín del mundo. Facultad de Estudios del Patrimonio Cultural de la Universidad Internacional SEK-Chile, Área de arqueología, Santiago de Chile, 169–180.

Schiavini A (1990) Estudio de la relación entre el hombre y los pinnípedos en el proceso adaptativo humano al canal Beagle, Tierra del Fuego, Argentina. PhD dissertation, Universidad de Buenos Aires, Buenos Aires

Schiavini A (1993) Los lobos marinos como recurso para cazadores-recolectores marinos: El caso de Tierra del Fuego. Lat Am Antiq 4(4):346–366

Schiavini A, Raya Rey A (2001) Aves y Mamíferos Marinos en Tierra del Fuego. Estado de situación, interacción con actividades humanas y recomendaciones para su manejo. Informe preparado bajo contrato con el proyecto Consolidación e Implementación del Plan de Manejo de la Zona costera Patagónica, proyecto ARG/97/G31 GEF/PNUD/MRECIC CADIC-CONICET

Sielfeld W (1983) Mamíferos Marinos de Chile. Ediciones de la Universidad de Chile, Santiago de Chile

Sielfeld W (1999) Estado del conocimiento sobre conservación y preservación de *Otaria flavescens* (Shaw, 1800) y *Arctocephalus australis* (Zimmermann, 1783) en las costas de Chile. Estud Oceanol 18:81–96

Tivoli A (2010a) Las aves en la organización socioeconómica de cazadores-recolectores-pescadores del extremo sur sudamericano. PhD dissertation, Universidad de Buenos Aires, Buenos Aires

Tivoli A (2010b) Temporal trends in avifaunal resource management by prehistoric sea nomads of the Beagle Channel region (southern South America). In: Prummel W, Zeiler J, Brinkhuizen D (eds) Birds in archaeology: proceedings of the 6th meeting of the ICAZ Bird Working Group in Groningen, Groningen archaeological studies, vol 10. Barkhuis Publishing, Groningen, pp 131–140

Tivoli A, Zangrando F (2011) Subsistence variations and landscape use among maritime hunter-gatherers. A zooarchaeological analysis from the Beagle Channel (Tierra del Fuego, Argentina). J Archaeol Sci 38(5):1148–1156

Winterhalder B (1986) Diet choice, risk, and food sharing in a stochastic environment. J Anthropol Archaeol 5(4):369–392

Zangrando A (2003) Ictioarqueología del canal Beagle. Explotación de peces y su implicación en la subsistencia humana. Sociedad Argentina de Antropología, Buenos Aires

Zangrando A (2009a) Historia evolutiva y subsistencia de cazadores-recolectores marítimos de Tierra del Fuego. Sociedad Argentina de Antropología: Colección Tesis de Doctorado, Buenos Aires

Zangrando A (2009b) Is fishing intensification a direct route to hunter-gatherer complexity? A case study from the Beagle Channel region (Tierra del Fuego, southern South America). World Archaeol 41(4):589–608

Zangrando A, Orquera L, Piana E (2010) Diversificación e intensificación de recursos animales en la secuencia arqueológica del canal Beagle (Tierra del Fuego, Argentina). In: Gutiérrez M, De Nigris M, Fernández P, Giardina M, Gil A, Izeta A, Neme G, Yacobaccio H (eds) Zooarqueología a principios del siglo XXI: aportes teóricos, metodológicos y casos de estudio. Ediciones del Espinillo, Buenos Aires, pp 359–370

Use of Marine Fauna and Tool Stones in the South of Buenos Aires Province (Argentina) During the Middle and Late Holocene

3

Romina Frontini and Cristina Bayón

3.1 Introduction

In order to characterize the exploitation of marine resources in the south of Buenos Aires province (Argentina) along the Holocene, we analyzed coastal and inland sites from a spatial and a temporal perspective. We considered two different lines of evidence: marine animal remains and lithic technology. Array analysis of these two discrete lines of evidence allowed complementing and enriching the inferences on the way of life of hunter-gatherers.

The Humid pampas of Buenos Aires were inhabited by hunter-gatherers from *ca.* 12,000 ^{14}C YBP (Politis 2008; Politis et al. 2014). Remains of these earliest settlements have been found inland (Politis and Bonomo 2011; Flegenheimer 2004; Martínez 2006; Mazzanti and Quintana 2001; Politis et al. 2012, 2014; among others), whereas the earliest remains found in Atlantic coastal sites date from *ca.* 7900 ^{14}C YBP (Bayón and Politis 2014).

The relationship between coastal and inland settlements has been widely studied. Researchers have focused on the chronology of settlements, and the movement between the coast and the continent (Bonomo 2005; and references therein). In the 1980s and 1990s, studies performed in the archaeological sites *La Olla* (sectors 1, 2, 3, and 4) and *Monte Hermoso* 1 (in the southwest coast of Buenos Aires province) provided the first radiocarbon dates for coastal settlements of hunter-gatherers. Studies in these sites also allowed recovering a great variety of archaeological evidence, including human footprints, faunal remains and a varied repertoire of technologies, the most prominent of which were wooden artifacts (Bayón and Politis 1996, 2014; Blasi et al. 2013; Johnson et al. 2000; Politis and Lozano 1988; Politis et al. 2009). These studies also highlighted the important role of

R. Frontini (✉) • C. Bayón
CONICET – Dpto. de Humanidades, Universidad Nacional del Sur, 12 de octubre y San Juan, 5to piso, 8000 Bahía Blanca, Buenos Aires Province, Argentina
e-mail: frontiniromina@gmail.com; crisbayon@gmail.com

M. Mondini et al. (eds.), *Zooarchaeology in the Neotropics*,
DOI 10.1007/978-3-319-57328-1_3

pinnipeds in the diet of hunter-gatherers. Two species of sea lion (*Otaria flavescens* and *Arctocephalus australis*) were abundantly represented and marine mollusks and continental fauna were also recovered (Johnson et al. 2000; Leon and Gutiérrez 2011). Regarding flaked rocks, quartzite was the predominant lithic raw material, followed by coastal pebbles. Both *La Olla* and *Monte Hermoso 1* are located in outcrops on the current beach resort of *Monte Hermoso*, adjacent to the sites on sand dunes described in this study. The chronology of all the settlements is coincident: *ca.* 7000 ^{14}C YBP.

Bonomo (2005) has recently analyzed coastal and inland sites between *Quequén Salado* river and *Mar del Plata* city and proposed a model of landscape use that suggests the complementary use of inland plains, sand dunes and coast from the Middle to the Late Holocene. The different kinds of settlements in each of these areas depended on the different activities carried out. Also, the archaeological site of Alfar, in *Mar del Plata* city (Buenos Aires province), indicates that by 5900 ^{14}C YBP, hunter-gatherers exploited marine resources, complementing them with continental mammals, mainly guanaco, and lithic artifacts manufactured on coastal pebbles (Bonomo and Leon 2010).

The use of coastal resources has also been studied through the analysis of stable isotopes of human remains from inland and coastal sites (Politis et al. 2009; Bonomo et al. 2013; Martínez et al. 2012).

Martínez (2008–2009) and Martínez et al. (2009, 2012) proposed a subsistence model that shows differences in the diet of hunter-gatherers that inhabited the *Colorado* River basin between the Middle Holocene (*ca.* 5000 ^{14}C YBP), the early Late Holocene (*ca.* 3000–1000 ^{14}C YBP) and the end of the Late Holocene (*ca.* 1000–250 ^{14}C YBP). The analysis of stable isotopes of the archaeological site *Cantera de Rodados Villalonga* suggested that during the Mid-Holocene, individuals had a marine diet and a mixed diet, with a strong marine component (Martínez et al. 2012). Based on the archaeofaunal record, Martínez (2008–2009) and Stoessel (2012) proposed that, at the end of the Late Holocene, there was diversification of the diet, based on the intensification of the use of faunal resources obtained from different environments, including the marine coast. During this period, there was an intensive use of fish, together with guanaco meat and vegetables (Stoessel 2012).

The aim of this study was to contribute to the regional knowledge of the way of life of hunter-gatherers during the Holocene, by highlighting tendencies of the marine resource exploitation in a region of the south of Buenos Aires province, Argentina. Also, we compared the new data and the available information from neighboring areas with the aim to contribute to better understanding the use of inland and coastal areas by hunter-gatherers in a regional scale.

3.2   Geographical and Environmental Settings

Buenos Aires province (34°–38°S; 56°–62°W) is located in the Humid Pampean region of Argentina. It is a flat grassland interrupted by the *Sierras Septentrionales* and *Sierras Australes* hilly systems, which are not higher than 1300 m a.s.l. To the east is the Atlantic Ocean, forming an extensive seashore that stretches over 1200 km (Cavallotto 2008). Within the apparent homogeneity, there are many environmental variations and thus different areas according to temperature, rainfall, topography, vegetation and fauna have been recognized (Politis 1984; Politis and Barros 2006; and references therein).

The study area (38°16′–38° 59′S; 60° 57′–62° 36′W) is located in a part of the South of *Buenos Aires* province in the grassland between the *Sierras Australes* and the sea (Fig. 3.1). Several water courses, such as *Sauce Grande* River, *Arroyo Napostá Grande*, and *Arroyo Napostá Chico*, flow from the *Sierras Australes* to the

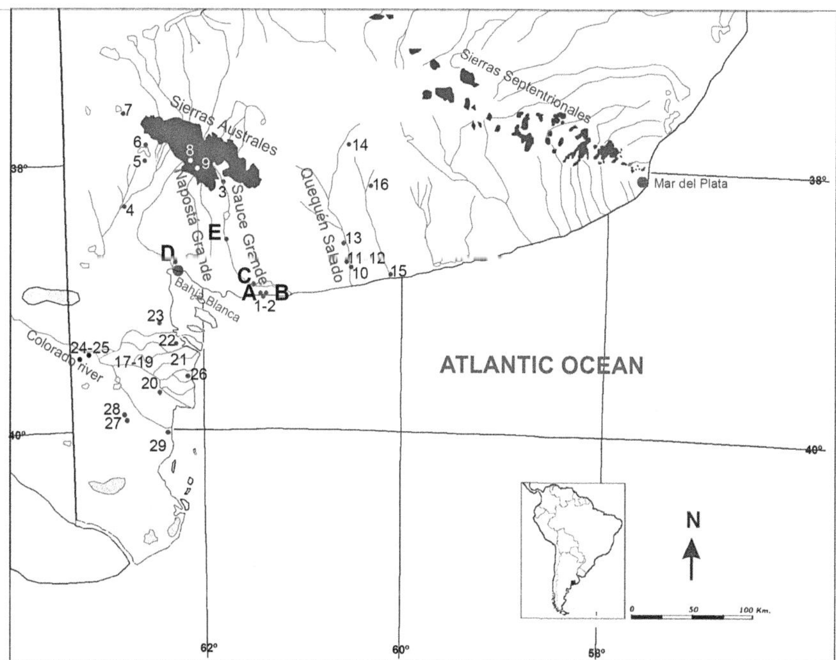

Fig. 3.1 Map showing the location of the archeological sites in study area, and sites considered in the discussion. A = El Americano II; B = Barrio Las Dunas; C = Puente de Fierro; D = Paso Vanoli; E = Paso Mayor. 1 = La Olla; 2 = Monte Hermoso 1; 3 = La Toma I; 4 = San Martín 1; 5 = Laguna Los Chilenos1; 6 = SA17 Avestruz; 7 = Laguna de Puán 1; 8 = El Abra; 9 = La Sofía 4; 10 = Quequén Salado 1; 11 = Quequén Salado 2; 12 = Quequén Salado 4; 13 = La Represa (QS7); 14 = Laguna Tres Reyes; 15 = Claromecó 1; 16 = Arroyo Seco 2; 17 = El Puma 2; 18 = El Puma 3; 19 = El Puma 4; 20 = El Tigre; 21 = La Petrona; 22 = La Primavera; 23 = Loma Ruiz 1; 24 = Paso Alsina 1; 25 = Zoko Andi; 26 = Localidad San Antonio; 27 = Loma de Los Morteros; 28 = La Modesta; 29 = Cantera de Rodados Villalonga

Atlantic Ocean. Environmentally, the study area is a broad transitional zone where ecotonal flora and fauna converge. For the purposes of our study, we characterized the availability of coastal and inland natural resources regarding fauna and lithic raw materials. Although the coast and the inland plains share several climatic and zoogeographic features, each environment has distinctive attributes that are worth mentioning from an archaeological perspective.

3.2.1 The Coast

In the study area, the Atlantic seashore has an East-West-Northwest direction and today presents sandy beaches, a cliff and the *Bahía Blanca* estuary (Cavallotto 2008; Fernández et al. 2003). During the Holocene, the beaches towards the east suffered significant geomorphological changes (Aramayo et al. 2005; Blasi et al. 2013; Quattrocchio et al. 2008). During the Early-Mid Holocene, near *Monte Hermoso* beach resort, the coast was associated with an estuary where diverse sub-environments were developed (Blasi et al. 2013). The coast with the current characteristics (i.e. sandy, open sea, and waves) was formed after 4000 ^{14}C YBP (Aramayo et al. 2005). Towards the continent, there is the *Barrera Medanosa Austral* (Isla et al. 2001; Monserrat 2010), a barrier of sand dunes that is about 4 km wide and was formed from the Middle Holocene.

3.2.1.1 Faunal Resources

In the coastal area, the most relevant taxa along the Holocene were pinnipeds (especially Southern sea lions and Southern fur seals) and sea fishes (e.g. *Pogonias cromis*; *Micropogonias furnieri*). Gastropods and bivalves (e.g. *Adelomelon* sp., *Zidona dufresnei*, and *Amiantis purpurata*), marine and terrestrial birds (*Larus* sp., *Phalacrocorax* sp., Ardeidae, *Rhea americana*, among others), and terrestrial mammals (i.e. tuco-tucos, armadillos, and foxes) were also present. Until the nineteenth century there were also guanacos and deer (Deschamps 2005).

3.2.1.2 Availability of Lithic Raw Material

The knappable rocks available in the study area are distinct types of pebbles and cobbles of different origin. The term "coastal pebbles" refers to a heterogeneous set of clasts transported by ocean currents from the mouth of North Patagonian great rivers (*Colorado* and *Negro*). They present great lithological variety, but igneous rocks (basalts, andesites, and rhyolites) are predominant, and tuffs and cherts are secondary. Coastal pebbles frequently range from 1 to 7 cm, and are scarce and scattered on the backshore (Bayón and Zavala 1997). Other rocks available in the study area are cobbles that have been transported from the *Sierras Australes* to the coast by the *Sauce Grande* River. Quartzites, quartz and subarkoses are the most abundant rocks. They are present only in the study area, spread along 13 km, between *Monte Hermoso* cliff and the beach resort of *Pehuen-có* (Bayón and Zavala 1997).

3.2.2 Inland Plains

The distinct orographic features of the inland plain are lakes, fluvial valleys, and interfluvial plains, generally with calcareous crusts. Paleoenvironmental studies from different proxies such as pollen, ostracods, vertebrates, and stratigraphy have shown climatic changes during the Holocene (Bayón and Zavala 1997; Deschamps 2005; Quattrocchio et al. 2008). While around 6000–5000 ^{14}C YBP, conditions were relatively humid, during the Late Holocene (*ca.* 2800 ^{14}C YBP), they were arid and semi-arid and the current climatic conditions were established (Quattrocchio et al. 2008).

3.2.2.1 Faunal Resources

The diversity of wildlife in the study area was higher in the past (Deschamps 2005). Large mammals, mainly *Lama guanicoe* and *Ozotoceros bezoarticus*, were especially relevant. Small mammals (e.g. *Ctenomys* sp.; Dasypodidae; *Pseudalopex* sp., *Lagostomus maximus*, *Dolichotis patachonica*, and *Conepatus* sp.), and birds (e.g., *Rhea americana*, Anatidae, etc.) were also present. The distribution of small mammals varied along the Holocene due to climatic changes and to the fact that the study area is ecotonal.

3.2.2.2 Availability of Lithic Raw Materials

The stone materials available in the inland plains are cobbles carried from the *Sierras Australes* to the coast by *Sauce Grande* River and *Arroyo Napostá Grande*. These materials are located in ancient terraces of fluvial valleys and are similar to those described for the coastal area.

The rocks include quartzites, quartz, subarkoses (Bayón et al. 1999, 2010; Vecchi 2011). Only some of these rocks were used for knapping, because fluvial cobbles have moderate or low quality. The size of fluvial cobbles varies significantly within the landscape (Bayón and Zavala 1997).

3.3 Materials and Methods

3.3.1 Archaeological Sites

The study area presents strong archaeological evidence of settlements in different environments (Bayón and Politis 2014; Bayón et al. 2006, 2010, 2012; Blasi et al. 2013; Frontini 2013; Frontini and Bayón 2015; Vecchi et al. 2013, 2014). In the present work, we considered five archaeological sites located in fluvial, wetland, and marine littoral environments: *Barrio Las Dunas* and *El Americano* II, which are in coastal landscape, and *Puente de Fierro*, *Paso Vanoli*, and *Paso Mayor*, which are located inland (Fig. 3.1; Table 3.1). For the purpose of this study, it is important to consider the distance from the coast of each site; coastal sites are located in the frontal sand dune while those further away are located between 7 and 41 km away (Table 3.1).

Table 3.1 Information on sites analyzed

Km to coast	Site	m^2	Chronol.	Enviro.	Total remains	Funct.	Ref.
0.65	*El Americano* II (EAII)	31	Middle-late Holocene	Coastal dunes	Lithic: 83 Faunal[a]: 421	Specific activities	This work
0.65	*Barrio Las Dunas* (BLD)	60	6924 ± 69 6.820 ± 100	Coastal dunes	Lithic: 894 Faunal[a]: 2132	Specific activities	Bayón et al. (2012)
7	*Puente de Fierro* (PF)	12	2000 ± 80 2042 ± 49	Wetland	Lithic: 257 Faunal[a]: 1368	Residential camp	This work; Frontini and Bayón (2015)
12.72	*Paso Vanoli* (PV)	20	630 ± 60 714 ± 53	Fluvial	Lithic: 454 Faunal[a]: 800	Residential camp and burials	Vecchi et al. (2013)
41	*Paso Mayor* YI S1 (lower component) (PM NI)	8	5877 ± 63 4046 ± 57 3820 ± 47	Fluvial	Lithic: 669 Faunal[a]: 2396	Residential camp	Bayón et al. (2010)
41	*Paso Mayor* YI S1 (upper component) (PM NS)	8	2774 ± 45 700 ± 42	Fluvial	Lithic: 463 Faunal[a]: 772	Residential camp and burial	Bayón et al. (2010) and Scabuzzo (2013)

[a]Faunal remains corresponds to Number of Specimens (NSP) comprising each assemblage

The temporal span represented is from 6900 to 700 ^{14}C YBP (Table 3.1). The sites considered include superficial as well as stratigraphic contexts. The three inland sites are stratigraphic contexts that were excavated by decapage, whereas in the coastal sites, the archaeological levels were exposed by Eolic dynamics. The field work in the coastal sites included mainly the systematic recovery of superficial remains and also, in the case of *El Americano* II, the excavation of a sub-superficial level up to 10 cm depth (Vecchi et al. 2014). Thus, the resolution and integrity of the contexts analyzed differ, and are particularly low in the coastal sites. Also, it is worth mentioning that sample sizes were affected by the postdepositional situations. Although *El Americano* II presents smaller sample size than the other sites (Table 3.1), we considered that its analysis was pertinent for the aim of the study as it a similar context to BLD site, and thus increase the information on sand dunes archaeofaunal assemblages. Based on radiocarbon dates and stratigraphic levels, *Paso Mayor* is a multicomponent site, where two components were identified (Bayón et al. 2010), and *Paso Vanoli* and *Puente de Fierro* are unicomponent. Regarding the coastal sites, they are palimpsests that probably included different chronology settlements along the Holocene. Due to the dynamic process of the dune environment, only the Middle Holocene levels preserved bone remains and it was thus possible to obtain radiocarbon dates. Regarding the Late Holocene, only stone tools were preserved and radiometric dates for these occupations could not be obtained.

In this study, we present original information of *El Americano* II and *Puente de Fierro*, where only a portion of the excavated area was considered. Information on the chronology, bioarchaeology, lithic technology, and zooarchaeology of *Paso Mayor*, *Paso Vanoli* and *Barrio Las Dunas* has been previously presented in several works (Bayón et al. 2006, 2010, 2012; Frontini 2013; Scabuzzo 2013; Vecchi 2011; Vecchi et al. 2013).

3.3.2 Analytical Methodology

This research was developed considering two types of evidence: archaeofaunal remains and lithic tool stones. The archaeofaunal remains were studied following the methodology proposed by several authors (Grayson 1984; Lyman 1994, 2008; Mengoni Goñalons 1999; among others). The faunal remains were anatomically and taxonomically identified; for comparison purposes, only the remains determined at least at Class level were considered. Otariid remains were determined by Dra. Florencia Borella. The quantification measures used to determine the taxonomic abundance were the number of identified specimens (NISP), the %NISP, and the minimum number of individuals (MNI) (Lyman 1994, 2008; Mengoni Goñalons 1999). MNI was obtained distinguishing right and left bones and differentiating between fused and unfused long bones (Mengoni Goñalons 1999). Measures of relative abundance of skeletal parts used were the minimum number of elements (MNE), the minimum number of anatomical units (MAU), and the %MAU. MNE was calculated based on the frequency at which each of the elements of the skeleton is represented, differentiating between fused and unfused bones (Mengoni Goñalons 1999). This parameter is linked to MAU, which was obtained by dividing the MNE of each anatomical unit by the number of times that this part is present in the entire skeleton. Its standardized measure (%MAU) allows examining the internal configuration of the assemblages.

Since the taphonomic features of each faunal context have been developed and presented in previous works (Bayón et al. 2010, 2012; Frontini and Bayón 2015; Frontini 2013; Vecchi et al. 2013, 2014), they will not be described here.

The human use of these faunal resources was evaluated based on the presence of cut marks, impact points, negative flake scars, fresh fractures and burnt bones (Mengoni Goñalons 1999; Lyman 1994). Transportation of resources was also considered. Finally, for each assemblage, the taxonomic abundance, the frequency and relative abundance of anatomical parts, the age classes, and the patterns of fracture were analyzed to evaluate human decisions in fauna handling (Grayson 1984; Lyman 1994; Mengoni Goñalons 1999).

Lithic assemblages were analyzed following the techno-morphological classification proposed by Aschero (1975, 1983) and Aschero and Hocsman (2004). According to the manufacturing technique, knap artifacts and grinding implements were sorted and for comparison purposes only flaked tools were considered. Cores, flakes, debris, and tools were differentiated according to the tool stone. Varieties of rocks were distinguished using the lithological repository of the *Universidad Nacional del Sur*, Bahía Blanca, Argentina. The relative abundance of each rock was analyzed to examine the areas of provenance of tool stones. Different types of cortex in flakes and cores were used to evaluate the presence of either coastal pebbles or river cobbles. Tool stones were grouped according to the probable source of supply. Coastal pebbles included basalts, andesites, rhyolites, and cherts, while river cobbles included quartzites, quartz, and subarkoses. Regarding the availability of raw materials in *Sierras Australes*, outcrops of quartzites, quartz and subarkose rocks were recorded, and the secondary deposits of river cobbles available in the study area came from this source. Thus, the presence of cobble cortex is useful to determine the probable tool stone provenance. Also rhyolites and muddy sandstone came from *Sierras Australes*, but these raw materials are not available in secondary deposits of river cobbles. Rocks transported from *Sierras Septentrionales* include mainly orthoquartzites from the *Sierras Bayas* group, orthoquartzites from the *Balcarce* Formation and cherts. *Sierras Septentrionales* were an important supply area. The sites studied are located more than 200 km away from this source. Other raw materials without known source were also studied.

3.4 Results

3.4.1 Zooarchaeological Record

Both marine and continental fauna (vertebrates and invertebrates) were found to be represented in four of the five sites analyzed, with different proportion according to the distance from the marine coast. In the coastal sites (*Barrio Las Dunas* and *El Americano* II), marine species were the most abundant, reaching between 85 and 94% NISP of the assemblages (Table 3.2). In contrast, in inland sites (7–41 km from the coast), continental species predominated while marine fauna represented only *ca.* 5%NISP of the assemblage, mainly in *Puente de Fierro* (Table 3.2). In *Paso Vanoli* and *Paso Mayor* YI S1 Upper levels, no marine fauna was recovered.

Among marine fauna, vertebrate (fish, birds, and mammals) as well as invertebrate (gastropods and bivalves) species were recovered (Table 3.2). Fish were represented only in both coastal sites. The species identified included *Pogonias cromis* (black drum), *Micropogonias furnieri* (croaker), and *Porichthys porosissimus* (catfish). Marine mammals included two otariid species: *Arctocephalus australis* (South American fur seal) and *Otaria flavescens* (South American sea lion), being the former the most abundant. Both were recorded in the two coastal sites and scarce elements were also represented in *Puente de Fierro*, 7 km from the coast. It is worth mentioning that in *Puente de Fierro* the otariid NISP was higher than in *El Americano* II (Table 3.2).

Marine invertebrate remains had been transported for great distances. They were recorded in both coastal and inland sites. Gastropod species included *Terebra*

Table 3.2 Marine species identified in the sites analyzed

	Coastal sites						Inland sites					
	El Americano II (NISP = 323)			*Barrio Las Dunas* (NISP = 968)			*Paso Mayor* NI (NISP = 178)			*Puente de Fierro* (NISP = 1362)		
Taxon	NISP	%NISP t	MNI	NISP	%NISP t	MNI	NISP	%NISP t	MNI	NISP	%NISP t	MNI
Otariidae	4	1.2	1	82	8.3	2	0	0	0	8	0.6	1
O. flavescens (South American sea lion)	0	0.0	0	1	0.1	1	0	0	0	1	0.1	1
A. australis (South American fur seal)	0	0.0	0	12	1.2	3	0	0	0	3	0.2	1
Teleostei	139	43	n/c	211	21.4	n/c	0	0	0	0	0.0	0
Scianidae	28	8.7	n/c	0	0	0	0	0	0	0	0.0	0
M. furneri (croaker)	18	5.6	8	1	0.1	1	0	0	0	0	0.0	0
P. cromis (black drum)	83	25.7	2	538	54.6	5	0	0	0	0	0.0	0
P. porostssimus (catfish)	0	0.0	0	1	6.1	1	0	0	0	0	0.0	0
Larus sp. (seagull)	0	0.0	0	0	0	0	0	0	0	2	0.1	2
Spheniscus sp. (penguin)	1	0.3	1	0	0	0	0	0	0	0	0.00	0
Total	**273**	**84.5**	**12**	**846**	**91.8**	**13**	**0**	**0**	**0**	**14**	**1**	**5**
Invertebrate	0	0	0	21	2.1	n/c	0	0	0	0	0	0
T. gennulata	0	0	0	0	0	0	0	0	0	1	0.1	1
Adelomelon sp.	0	0	0	0	0	0	1	0.5	1	1	0.1	1
A. brasiliana	0	0	0	0	0	0	0	0	0	3	0.2	3
Z. dufresnei	0	0	0	0	0	0	1	0.5	1	0	0.0	0
A. purpurata	0	0	0	0	0	0	1	0.5	1	1	0.1	1
M. patachonica	0	0	0	0	0	0	0	0	0	3	0.2	2
Mactra sp.	0	0	0	0	0	0	0	0	0	45	3.3	–
Total	**0**	**0**	**0**	**21**	**2.1**	**0**	**3**	**1.5**	**3**	**54**	**4**	**8**

NISP, MNI, and %NISP t = proportional NISP obtained from the total Number of Identified Specimens (NISP), *n/c* does not correspond *gennulata, Adelomelon* sp., *Adelomelon brasiliana,* and *Zidona dufresnei,* whereas bivalves included *Amiantis purpurata, Mactra* sp., and *Mactra patachonica.* Among the sites studied, *Puente de Fierro* recorded the greatest diversity and abundance of invertebrates (Table 3.2). There, four different species (which represents *ca.* 4% total NISP) were identified. The presence of a bead was also recorded, which indicates technological use.

Regarding the information of the anatomical representation of the best represented species, head bones of fish as well as postcranial (axial and appendicular) elements were recovered, suggesting that they were transported whole to both sites. However, the %MAU values indicated some differences with respect to their degree of representation in each site (Fig. 3.2). In *Barrio Las Dunas,* the postcranium was the anatomical section best represented by the dorsal and anal fin first pterygiophores, whereas facial bones and opercularis had a low representation and were the only head elements identified. Neurocranium elements were absent. Conversely, in *El Americano* II, both the cranial and postcranial skeleton had a high representation according to %MAU. The ceratohyal was the most highly represented element. The neurocranium was well represented by frontal and supraoccipital bones (*ca.* 65%MAU), whereas the operculum, premaxilla, and quadrate were absent. The postcranium was well represented by anal fin first pterygiophores and precaudal vertebrae. The differences in the representation of anatomical elements in both sites could be due to cultural or natural causes. Post-depositional effects could have differentially affected both contexts, mainly considering that sand dunes are highly dynamic environments, where the re-exposure and reburial of archaeological deposits are common processes. The survival potential of different bones varies markedly between fish species. Also the differential

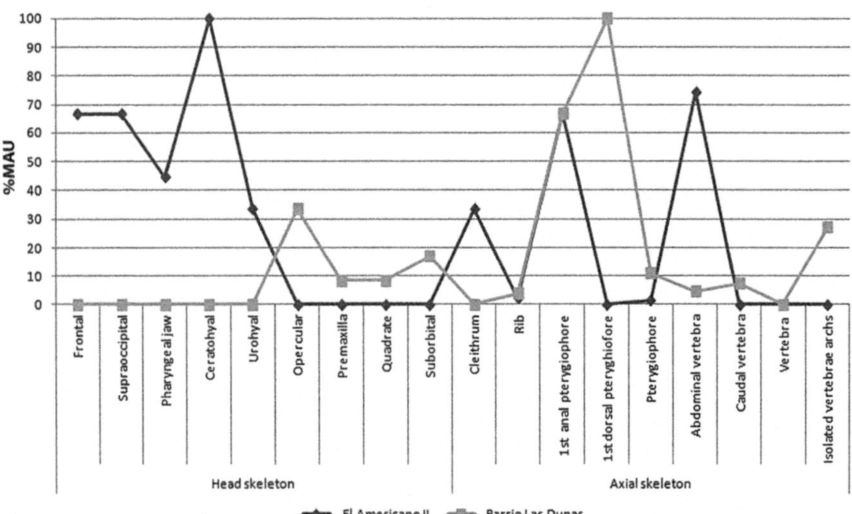

Fig. 3.2 *Pogonias cromis* anatomical representation (%MAU) in *El Americano* II (NISP = 83) and Barrio Las Dunas (NISP = 538) sites

bone density in different elements of fish skeleton is relevant. As taphonomic studies have been developed on fish species other than the black drum (i.e. salmonids), with smaller size, in fluvial contexts and seashore (Butler 1993; Falabella et al. 1994; Lubisnki 1996), the results of these studies are not applicable to our study case. Undoubtedly, further actualistic and experimental work is needed to gain better understanding of how different skeletal elements are affected by post-depositional sand dune processes.

Regarding cultural causes, the differences observed in the representation of anatomical elements could derive from differences in fish processing, which included differential treatment of heads and trunks because of differences in the distribution of soft tissues along fish anatomy (Butler 1993; Stewart and Gifford-González 1994; Zohar et al. 2001). Fish head processing included smashing the head to extract the brain, as documented by Stewart and Gifford-González (1994). This would lead to a great fragmentation or destruction of head bones, a pattern that may be coincident with the scarce head elements recovered in *Barrio Las Dunas*.

Although samples of marine mammals in *El Americano* II and *Puente de Fierro* were small, the elements recovered showed a similar pattern in *Barrio Las Dunas*, *El Americano* II and *Puente de Fierro* (Table 3.3; Fig. 3.3) because, in the three contexts, forelimbs and hind limbs were the most common anatomical parts. In *Barrio Las Dunas*, also axial and cranial elements were recovered (Fig. 3.3). The representation of anatomical parts in this site could be explained by the income of whole carcasses to the site (Bayón et al. 2012). This would be probable as the site was located near the provisioning places.

Evidence of human use of marine resources was recorded in *Pogonias cromis*, Otariidae and invertebrates. Burnt bones and scarce cut marks were identified in *P. cromis* and otariids (*Barrio Las Dunas* and *El Americano* II). Appendicular elements of otariids had been carried from the marine coast to 7 km inland to *Puente de Fierro* and gastropods and bivalves had been carried for more than 40 km (*Paso Mayor* Lower Component).

Table 3.3 NISP values of Otariidae elements represented in El Americano II and Puente de Fierro

| | *El Americano* II (NISP = 4) | | | *Puente de Fierro* (NISP = 12) | | | |
| | Unfused | | | Unfused | | | |
Element	L	R	Undet	L	R	Undet	Axial
Postcanine							1
Humerus		1		1	1		
Ulna	1				2		
Ileon					1		
Femur		1					
Metapodial				1		2	
Astragalus						1	
Phalanx			1			2	

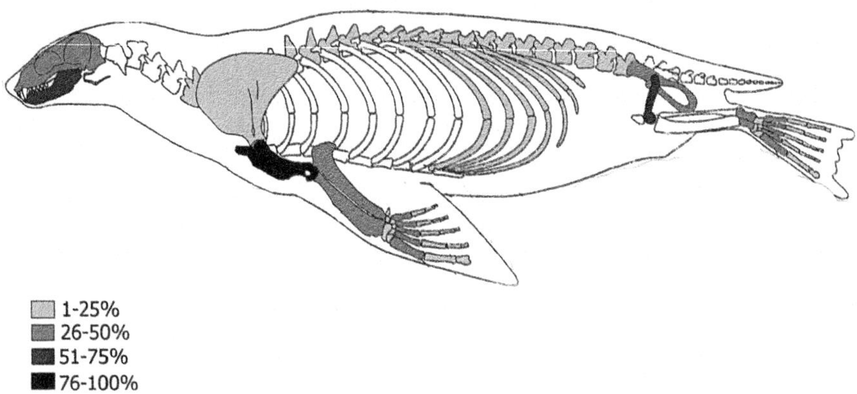

Fig. 3.3 %MAU anatomical representation of Otariids in *Barrio Las Dunas* site. (Taken from Bayón et al. 2010)

3.4.2 Lithic Assemblages

Regarding the procurement of tool stones by hunter-gatherers in the study area, rocks from different sources were present in the five sites analyzed although in different proportions, because the local availability of lithic resources influenced the composition of artifact assemblages (Table 3.4). Regarding coastal pebbles, they were identified in the five archaeological sites studied. Their transport along different distances and their use allow discussing similarities and differences between sites. In coastal sites, coastal pebbles were used as nodules to knap. In both sites, they were the most common tool stone flaked; at least 50% of the instruments were made on these pebbles. Coastal pebbles also predominated among cores and debitage (77%). Both assemblages showed evidence that the most frequent type of reduction was bipolar knapping, followed by free hand knapping. Diverse kinds of tools such as end scrapers, side scrapers, denticulates and expediently retouched flakes, were found. Most of these tools were complete. The tools found showed little modification and marginal retouch. Bifacial retouch was scarce and only fragmented bifaces, particularly two projectile points, were recovered in these sites.

In inland sites, coastal pebbles were used in variable proportions (Table 3.4). We found inter-assemblage variability with different proportions of tool stones and manufacturing techniques in each site. Different trajectories of stone tool production were present depending on local availability, distance of transport and rock quality, among others. In *Puente de Fierro* and *Paso Vanoli* (7 and 13 km from the coast, respectively), there were no raw materials immediately adjacent to the sites. In both cases, all tool stones were transported and coastal pebbles were the first stone of choice to make tools, while other secondary rocks were transported from *Sierras Septentrionales*, fluvial cobble deposits of *Sauce Grande* River, and *Sierras Australes*. We also found rocks of unknown sources. The assemblages presented a

Table 3.4 Information on tool stones according to artifacts type in each site

Site	Artifacts	Coastal pebbles	Fluvial cobbles	*Sierras Australes*	*Sierras Septentrionales*	Other rocks
El Americano II (*n* = 83)	Tools	1	0	0	1	0
	Cores	7	2	0	0	0
	Flakes and debris	56	10	3	2	1
	Total	**64**	**12**	**3**	**3**	**1**
Barrio Las Dunas (*n* = 894)	Tools	20	7	0	7	1
	Cores	57	5	0	0	0
	Flakes and debris	638	113	1	33	12
	Total	**715**	**125**	**1**	**40**	**13**
Puente de Fierro (*n* = 257)	Tools	21	9	1	18	0
	Cores	8	2	0	5	2
	Flakes and debris	69	34	5	73	10
	Total	**98**	**45**	**6**	**96**	**12**
Paso Vanoli (*n* = 422)	Tools	9	3	3	1	0
	Cores	7	3	0	1	0
	Flakes and debris	94	210	14	54	23
	Total	**110**	**216**	**17**	**56**	**23**
Paso Mayor NI (*n* = 669)	Tools	3	36	3	18	2
	Cores	3	28	0	3	0
	Flakes and debris	22	446	27	50	28
	Total	**28**	**510**	**30**	**71**	**30**
Paso Mayor NS (*n* = 463)	Tools	7	21	4	6	1
	Cores	3	9	0	3	0
	Flakes and debris	18	310	11	39	31
	Total	**28**	**340**	**15**	**48**	**32**

variety of tool types, including end scrapers, side scrapers, borers, projectile points, denticulates and retouched flakes. Tools were made on flakes, and most of the artifacts were flaked by unifacial or bifacial retouch. End scrapers, which were made on coastal pebble flakes, were the most frequent artifacts.

In contrast, in *Paso Mayor* (41 km from the coast), we found locally available fluvial cobbles of variable knapping quality (regular to poor, depending on the frequency of fissures and on the degree of weathering). These fluvial cobbles were used in an expedient way to manufacture mostly unifacial and marginal tools and a wide range of retouched flake tools, which were typologically unidentified. Lithic production in *Paso Mayor* was an important activity and debitage frequency increased because lithic raw materials are abundant. Coastal pebbles represented less than 6% in total. In Late Holocene settlements, they constituted the second rock in importance (17.5%) for making artifacts. However, the underrepresentation of cores and debitage point out that they were probably transported as tools. Flakes of coastal pebbles were used mainly to manufacture projectile points and end scrapers. In this site, although the importance of coastal pebbles varied along the Holocene, it seems to be related to the continuity of transport from the coast to the site as tool stones.

3.5   Discussion

The analysis of two material-type evidences—faunal remains and stone tools—in the archaeological sites from the study area allowed reconstructing temporal and spatial patterns in the exploitation of these marine resources. We were able to recognize temporal continuities and changes along the Holocene. The temporal continuity was evidenced by the recurrent use of marine mollusks, as they were found in four of the five sites studied, beyond their chronology. The other evidence of temporal continuity was the presence of coastal pebbles, which were recovered in the five sites analyzed in percentages ranging from 5 to 77%.

The temporal difference recognized is related to the exploitation of marine fauna. In the Middle Holocene, marine vertebrates were preferentially used. In these times, fishes (mainly black drums) were obtained in the paleo-estuarine coast and processed in the sand dunes (*Barrio Las Dunas* and *El Americano* II). In less proportion, pinnipeds of two species were found on the beach. These were butchered and consumed in different sectors of the ancient estuary (Blasi et al. 2013). The archaeological site *La Olla* (sectors 1, 2, 3, and 4), which has a chronology similar to that of the coastal sites studied, also indicates a great importance of marine resources, although pinnipeds have higher representation than fishes (Bayón and Politis 2014; Johnson et al. 2000; Leon and Gutiérrez 2011). In contrast, Late Holocene archaeological assemblages showing the use of marine vertebrates are exiguous, being that scarce remains of pinnipeds were recovered only in *Puente de Fierro* (*ca.* 2000 14C YBP). If this situation is not affected by preservation biases, it could be stated that the exploitation of marine vertebrates decreased during the Late Holocene. This would indicate cultural decisions on the selection of resources. Also, this could be explained due to a reduction in the local availability of marine mammals and fish in the Atlantic coast. The paleoenvironmental reconstruction shows that the current sandy coast was established after *ca.* 4000 YBP (Quattrocchio et al. 2008). Certainly, this could have affected the distribution of Middle Holocene marine fauna.

The results obtained by Politis et al. (2009), Bonomo et al. (2013), and Martínez et al. (2012) indicate that the consumption of marine resources in the *Interserrana* plain and *Colorado* River basin varied temporally along the Holocene. At the end of the Early Holocene and during the Middle Holocene, there was an important use of marine resources, reflected in enriched isotopic values that indicate marine and mixed diets. During the Late Holocene, the isotopic values indicate mixed diets, mainly based on terrestrial species and, complementarily, on few marine resources. Thus, isotopic analyses are concordant with the proposal of the decrease in the consumption of marine resources during the Late Holocene.

Regarding the spatial pattern, we identified differences between coastal and inland archaeological contexts, consistent with the temporal tendencies. The coastal sites indicated, with a strong archaeological signal, the importance of the use of marine resources (fauna and coastal pebbles), whereas inland sites showed that only tool stones and mollusks were transported. Marine vertebrate remains were absent in most of the inland sites. In the coastal sites, coastal pebbles were the first raw material selected, and these were flaked by the bipolar technique (Bayón et al. 2012; Flegenheimer et al. 1995; Vecchi et al. 2014). Also, otariids and fish were obtained and processed near the provisioning places. Inland, the tool stones gathered in the coast were relevant, considering that coastal pebbles constituted the first option to knap tools in both *Puente de Fierro* and *Paso Vanoli* (7 and 13 km from the coast). From 5800 YBP to 700 YBP, transportation was also relevant in *Paso Mayor*, 41 km from the coast (Bayón et al. 2010).

The archaeological information of the assemblages from the study area indicates the importance of the coast as a focus of resource acquisition all along the Holocene. We can state that the importance of marine food was reduced in the Late Holocene, but that the importance of coastal tool stones and mollusks continued over time.

By comparing the patterns of the sites studied with those of sites located *ca.* 150 km around the coast of our study area (Table 3.5), we found the following. Regarding faunal remains, marine vertebrates, including mammals and fishes, were present in four of the contexts studied. In *La Olla*, pinnipeds were highly represented in the Mid-Holocene, whereas *in San Antonio*, marine fishes were highly represented in the Late Holocene (Table 3.5). Marine mollusks showed a different situation, as they were present in 13 of the 29 contexts analyzed, reaching distances of 90 km from the coast. Although the relative importance in each assemblage is low, there is a great variety of species represented, including *Adelomelon brasiliana, Adelomelon beckii, Amiantis purpurata, Pecten* sp., among others. Also, there are evidences of technological use of mollusks as raw material to manufacture beads, mainly related to human burials. The transportation along hundreds of kilometers, the low frequency and the great taxonomic diversity of marine mollusks are coincident with that proposed by Bonomo (2007), who analyzed 20 settlements in the Dry Pampa, the Humid Pampa and North Patagonia. This author also proposed that, for the Pampean region, the use of marine mollusks was related to the manufacture of personal ornaments, which acted as non-verbal communication stuff.

The use of tool stones shows diverse patterns in the different adjacent areas. Regarding rock provisioning in the *Interserrana* area, Bonomo (2005) proposed that in the sites located on sand dunes and between 3 and 11 km from the seashore, coastal pebbles were the main rocks chosen for knapping. This preference for coastal pebbles is related to the areas of natural availability. While moving away from the coast, the use of other rocks increased progressively. These raw materials,

Table 3.5 Information of marine resource exploitation from sites located at *ca.*150 km around the coast of our study area

No	Site	^{14}C YB	Km from current coast	Fish	Mam.	Inver.	Coastal pebbles
MH	*Arroyo Seco* (Politis et al. 2014)	8460 ± 74 to 5793 ± 64	60	–	–	X	X
	La Olla (1–4) (Leon and Gutiérrez 2011)	*ca.* 7900–6700	0	X	X	X	X
	Monte Hermoso 1 (Bayón and Politis 2014)	*ca.* 7900–6700	0	–	–	X	X
	El Abra (Castro 1983)	6230 ± 90	90	–	–	–	–
	La Modesta (Stoessel 2015)	5641 ± 66	60	–	–	–	N/A
	Loma de los Morteros (Stoessel 2015)	4454 ± 60 4269 ± 59	55	–	–	–	N/A
	Cantera de Rodados Villalonga (Martínez et al. 2012)	4889 ± 58 to 4100 ± 80	8	–	–	–	N/A
Early LH	SA 17 Avestruz (Austral 1994)	3280 ± 70	90	–	–	X	N/I
	Laguna de Puan 1 (Oliva et al. 1991b)	3300 ± 100	100	–	–	–	–
	La Represa (QS7) (Hoguin and March 2007–2008)	3430 ± 40 to 2110 ± 40	23	–	–	–	X
	La Primavera (Martínez 2008–2009)	*ca.* 3000–2800 años	20	X	–	X	N/A
	San Martin 1 (Oliva et al. 1991a)	2890–80	80	–	–	X[a]	X
	Laguna Tres Reyes (Madrid and Barrientos 2000)	2470 ± 60 to 1845 ± 50	110	–	–	–	X
	El Puma 3 (Martínez et al. 2012)	2209 ± 48	80	–	–	–	N/A
Early LH	*La Toma I* (CS) (Salemme 1987; Álvarez 2012)	2075 ± 70 to 995 ± 65	40	–	–	X	X
	Loma Ruiz 1 (Martínez and Martínez 2011)	1935 ± 44 to 1615 ± 50	28	–	–	–	N/A
	Quequén Salado 2 (Madrid et al. 2002)	1720 ± 40	11	–	–	–	–
	El Puma 4 (Martínez and Martínez 2011)	1832 ± 51	80	–	–	–	N/A
	La Sofia 4 (Oliva 2000)	1595 ± 70	120	–	–	–	–
	El Puma 2 (Martínez and Martínez 2011)	1548 ± 51	80	–	–	–	N/A
	Zoko Andi (Martínez et al. 2014)	1527 ± 34 to 380 ± 43	80	–	–	–	N/A

(continued)

Table 3.5 (continued)

No	Site	^{14}C YB	Km from current coast	Fish	Mam.	Inver.	Coastal pebbles
	Quequén Salado 4 (Madrid et al. 2002)	1240 ± 40	11	–	–	–	–
Final LH	*Localidad San Antonio* (Martínez 2008–2009)	1053 to 764 ± 45	5	X	–	X	N/A
	Quequén Salado 1 (Leon 2014)	960 ± 40 to 360 ± 40	11	–	–	–	X
	El Tigre (Martínez 2008–2009)	930 ± 47 to 437 ± 43	20	X	–	–	N/A
	Claromecó 1 (Bonomo et al. 2008; Leon 2014)	800 ± 34	3	–	–	X	X
	La Petrona (Martínez 2004)	770 ± 49 to 248 ± 39	61	–	–	X	N/A
	Paso Alsina 1 (Martínez et al. 2007)	570 ± 44 to 448 ± 43	100	–	–	X	N/A
	Laguna Los Chilenos 1 (Barrientos 1997)	470 ± 40	90	–	–	X[a]	–

References: *MH* middle-Holocene, *LH* late Holocene, *Mam* mammalian, *Inver* invertebrate, *X* exploitation, *N/I* no information, *N/A* not applicable
[a]Technological use

in particular orthoquartzites of the *Sierras Bayas* Group, came from *Sierras Septentrionales*. However, coastal pebbles as secondary lithic resources were transported to sites located 40, 60, and 190 km from the seashore. The use of coastal pebbles in the *Interserrana* area and that in the study area present many similarities. The presence of coastal pebbles in inland sites and their prevalence in coastal sites were recorded in both areas, and bipolar flaking was part of the tool manufacturing choices. The main difference is related to the natural availability, as coastal pebbles were more abundant, even currently, in the *Interserrana* plain coast than in the study area (Bonomo 2005).

In sites located in the *Colorado* River basin, local raw materials (siliceous rocks, chalcedony, and basalts), locally named *rodados Tehuelches*, are predominant in archaeological assemblages. These small pebbles are abundant in the landscape and available throughout the coastal and inland areas. They were knapped to make tools, as the most important tool stone selected (Armentano 2012; Martínez 2008–2009). Bipolar flaking was part of the tool manufacturing repertoire. The main difference is the great availability of pebbles in both inland and coastal sites. Similarities between the two areas, such as the choice of bipolar flaking to knap small pebbles, were observed.

In summary, in the study area, during the Middle Holocene, there was a strong archaeological signal of the use of both vertebrate and invertebrate marine fauna and coastal pebbles in coastal sites. In inland sites, mollusks and coastal pebbles were transported at least 41 km from coast. By the Late Holocene, marine invertebrates and coastal pebbles were relevant in inland sites, and scarce pinniped remains were recorded 7 km from the coast. It can be stated that the importance of marine food decreased over time and that the importance of coastal stone tools and mollusks continued over time as it is evidenced by inland sites.

Based on the two discrete lines of evidences studied, faunal remains and lithic raw material, we can state that the Atlantic coast was regularly visited by hunter-gatherers all along the Holocene. For thousands of years, social decisions on feeding, technological use, and ideological activities have underlain the relative importance that hunter-gatherers gave to the different marine supplies such as lithic raw materials, mollusks, and vertebrates.

Acknowledgments A shorter version of this manuscript was originally presented at the Neotropical Zooarchaeology Session at the 12th ICAZ International Conference (2014). Rodrigo Vecchi and Clara Scabuzzo made valuable comments to a first draft. Projects BID—PICT 2013-0179 and SECYT-UNS 24/I 182 supported this research. Two anonymous reviewers made valuable comments that helped to clarify the ideas expressed and improve the manuscript. The authors are the sole responsible for the ideas expressed herein.

References

Álvarez MC (2012) Análisis Zooarqueológicos en el sudeste de la región pampeana. Patrones de subsistencia durante el Holoceno tardío. Unpublished PhD thesis. Facultad de Ciencias Sociales, Universidad Nacional del Centro de la Provincia de Buenos Aires, Olavarría

Aramayo SA, Gutiérrez de Téllez B, Schillizzi RA (2005) Sedimentologic and palaeontologic studies of the southeast coast of Buenos Aires province, Argentina: a late Pleistocene-Holocene paleoenvironmental reconstruction. J S Am Earth Sci 20:65–71

Armentano G (2012) Arqueología del curso inferior del Río Colorado. Estudio tecnológico de las colecciones líticas de Norpatagonia Oriental durante el Holoceno tardío. Departamentos de Villarino y Patagones, Provincia de Buenos Aires, Argentina. Unpublished PhD thesis. Universidad Nacional del Centro de la Provincia de Buenos Aires and Université de Paris OuestNanterre-La Défense, Olavarría and Nanterre

Aschero C (1975) Ensayo para una clasificación morfológica de artefactos líticos aplicada a estudios tipológicos comparativos. CONICET, Buenos Aires

Aschero C (1983) Ensayo para una clasificación morfológica de artefactos líticos aplicada a estudios tipológicos comparativos. Revisión. CONICET, Buenos Aires

Aschero C, Hocsman S (2004) Revisando cuestiones tipológicas en torno a la clasificación de artefactos bifaciales. In: Acosta et al (eds) Temas de Arqueología, Análisis Lítico. Universidad Nacional de Luján, Luján, pp 7–25

Austral A (1994) Arqueología en el Sudeste de la Provincia de Buenos Aires. Actas y memorias del XI Congreso Nacional de Arqueología Argentina (Resúmenes). Revista del Museo de Historia Natural de San Rafael 14:201–204

Barrientos G (1997) Nutrición y dieta de las poblaciones aborígenes prehispánicas del sudeste de la región pampeana. Unpublished PhD thesis, Facultad de Ciencias Naturales y Museo, Universidad Nacional de La Plata, La Plata

Bayón C, Politis G (1996) Estado actual de las investigaciones en el sitio Monte Hermoso I (Prov. de Buenos Aires). Arqueología 6:83–116

Bayón C, Politis G (2014) The inter-tidal zone site of La Olla: early–middle Holocene human adaptation on the Pampean Coast of Argentina. In: Evans AM et al (eds) Prehistoric archaeology on the continental shelf. Springer, New York, pp 115–130

Bayón C, Zavala C (1997) Coastal sites in southern Buenos Aires: a review of 'Piedras Quebradas'. In: Rabassa J, Salemme M (eds) Quaternary of South America and Antarctic Peninsula. A.A. Balkema, Rotterdam, pp 229–254

Bayón C, Flegenheimer N, Valente M et al (1999) Dime cómo eres y te diré de dónde vienes: la procedencia de rocas cuarcíticas en la Región Pampeana. Relac Soc Arg Antr 24:187–232

Bayón C, Flegenheimer N, Pupio A (2006) Planes sociales en el abastecimiento y traslado de roca en la Pampa bonaerense en el Holoceno temprano y tardío. Relac Soc Arg Antr 31:19–45

Bayón C, Pupio A, Frontini R et al (2010) Localidad Arqueológica Paso Mayor: nuevos estudios 40 años después. Intersec Antr 11:155–166

Bayón C, Frontini R, Vecchi R (2012) Middle Holocene settlements in coastal dunes from southwest of Buenos Aires province, Argentina. Quat Int 256:54–61

Blasi A, Politis G, Bayón C (2013) Palaeoenvironmental reconstruction of La Olla, a Holocene archaeological site in the Pampean coast (Argentina). J Archaeol Sci 40:1554–1567

Bonomo M (2005) Costeando las llanuras. Arqueología del litoral marítimo pampeano. Sociedad Argentina de Antropología, Buenos Aires

Bonomo M (2007) El uso de los moluscos marinos por los cazadores-recolectores pampeanos. Chungara, Revista de Antropología Chilena 39(1):87–102

Bonomo M, Leon DC (2010) Un contexto arqueológico en posición estratigráfica en los médanos litorales. El sitio Alfar (Pdo. de General Pueyrredón, Pcia. De Buenos Aires). In: Berón M et al (eds) Mamul Mapü: pasado y presente desde la arqueología pampeana, II. Libros del Espinillo, Ayacucho, pp 29–45

Bonomo M, Leon DC, Turnes L et al (2008) Nuevas investigaciones sobre la ocupación prehispánica de la costa pampeana en el Holoceno tardío: el sitio arqueológico Claromecó 1 (partido de Tres Arroyos, provincia de Buenos Aires). Inter Antr 9:24–41

Bonomo M, Scabuzzo C, Leon DC (2013) Cronología y dieta en la costa atlántica pampeana, Argentina. Inter Antr 14:123–136

Butler V (1993) Natural versus cultural salmonid remains: origins of the Dalles Roadcut bones, Columbia River, Oregon, U.S.A. J Archaeol Sci 20:1–24

Castro A (1983) Noticia preliminar sobre un yacimiento en la Sierra de la Ventana, Sierras Australes de la Provincia de Buenos Aires. Relac Soc Arg Antr 15:91–107

Cavallotto JL (2008) Geología y geomorfología de los ambientes costeros y marinos. In: Atlas de Sensibilidad Ambiental de la Costa y el Mar Argentino. http://atlas.ambiente.gov.ar/tematicas/mt_01/geologia.htm. Accessed Feb 2015

Deschamps CM (2005) Late Cenozoic mammal bio-chronostratigraphy in southwestern Buenos Aires Province, Argentina. Ameg 42(4):733–750

Falabella F, Loreo Vargas M, Meléndez R (1994) Differential preservation and recovery of fish remains in Central Chile. In: van Neer W (ed) Annales du Musée Royal de l'Afrique Centrale, Sciencias Zoologiques, vol 274, pp 25–35

Fernández E, Caló J, Marcos A, Aldacour H (2003) Interrelación de los ambientes eólico y marino a través del análisis textural y mineralógico de las arenas de Monte Hermoso, Argentina. Rev Asoc Arg Sed 10(2):151–161

Flegenheimer N (2004) Las ocupaciones de la transición Pleistoceno-Holoceno: una visión sobre las investigaciones en los últimos 20 años en la Región Pampeana. In: Beovide L, Barreto I, Curbelo C (eds) La Arqueología uruguaya ante los desafíos del nuevo siglo. Montevideo, Uruguay, Asociación Uruguaya de Arqueología, pp 2–17

Flegenheimer N, Bayón C, González de Bonaveri MI (1995) Técnica simple, comportamientos complejos: La talla bipolar en la arqueología bonaerense. Relac Soc Arg Antr 20:81–110

Frontini R (2013) Aprovechamiento faunístico en entornos acuáticos del sudoeste bonaerense durante el Holoceno (6900-700 años AP). Relac Soc Arg Antr 38(2):493–519

Frontini R, Bayón C (2015) Consumo de recursos animales de porte menor durante el Holoceno tardío en el sudoeste de la provincia de Buenos Aires (Argentina). Arch Inter J Archae 24:271–293

Grayson D (1984) Quantitative zooarchaeology: topics in the analysis of archaeological faunas. Academic Press, Orlando

Hoguin R, March R (2007–2008) Una primera aproximación al análisis tipo-tecnológico de los artefactos líticos del sitio La Represa (curso inferior del Quequén Salado, provincia de Buenos Aires). Arqueología 14:103–136

Isla FI, Cortizo LC, Turno Orellano HA (2001) Dinámica y evolución de las barreras medanosas, Provincia de Buenos Aires, Argentina. Rev Bra Geomor 2(1):73–83

Johnson E, Politis G, Gutiérrez MA (2000) Early Holocene bone technology at the La Olla 1 site, Atlantic Coast of the Argentine Pampas. J Archaeol Sci 27(6):463–477

Leon DC (2014) Zooarqueología de cazadores recolectores del litoral pampeano: un enfoque multidimensional. Unpublished PhD thesis, Universidad Nacional del Centro de la Provincia de Buenos Aires, Olavarría

Leon C, Gutiérrez MA (2011) Análisis faunístico de los sectores 3 y 4 del sitio La Olla (LO3 y LO4) en el litoral Atlántico bonaerense. In: Libro de Resúmenes del VI Congreso de Arqueología de la Región Pampeana Argentina, La Plata

Lubisnki PM (1996) Fish heads, fish heads: an experiment on differential bone preservation in a Salmonid fish. J Archaeol Sci 23:175–181

Lyman RL (1994) Vertebrate taphonmy. Cambridge University Press, London

Lyman RL (2008) Quantitative paleozoology. Cambridge University Press, London

Madrid P, Barrientos G (2000) La estructura del registro arqueológico del sitio Laguna Tres Reyes (Provincia de Buenos Aires): Nuevos datos para la interpretación del poblamiento humano del sudeste de la región pampeana a inicios del Holoceno Tardío. Relac Soc Arg Antr 25:179–206

Madrid P, Politis G, March R et al (2002) Arqueología microrregional en el sudeste de la región pampeana argentina: el curso del río Quequén Salado. Relac Soc Arg Antr 27:327–355

Martínez G (2004) Resultados preliminares de las investigaciones arqueológicas realizadas en el curso inferior del río Colorado (Partidos de Villarino y Patagones, Provincia de Buenos Aires). In: Martínez G, Gutiérrez M, Curtoni R et al (eds) Aproximaciones Contemporáneas a la Arqueología Pampeana. Perspectivas Teóricas, Metodológicas, Analíticas y Casos de Estudio. Facultad de Ciencias Sociales, Olavarría, pp 275–292

Martínez G (2006) Arqueología del curso medio del río Quequén Grande: estado actual y aportes a la arqueología de la región pampeana. Relac Soc Arg Antr 31:249–275

Martínez G (2008–2009) Arqueología del curso inferior del río Colorado: estado actual del conocimiento e implicaciones para la dinámica poblacional de cazadores recolectores pampeano-patagónicos. Caz Rec Con Sur 3:71–92

Martínez G, Martínez GA (2011) Late Holocene environmental dynamics in fluvial and aeolian depositional settings: archaeological record variability at the lower basin of the Colorado River (Argentina). Quat Int 245:89–102

Martínez G, Flensborg G, Bayala P et al (2007) Análisis de la composición anatómica, sexo y edad de los entierros secundarios del sitio Paso Alsina 1 (Pdo Patagones, Buenos Aires). In: Bayón C, Flegenheimer N, González de Bonaveri MI et al (eds) Arqueología en las Pampas. Sociedad Argentina de Antropología, Bahía Blanca, pp 41–58

Martínez G, Zangrando AF, Prates L (2009) Isotopic ecology and human paleodiets in the lower basin of the Colorado River (Buenos Aires province, Argentina). Int J Osteoarchaeol 19:281–296

Martínez G, Flensborg G, Bayala PD (2012) Primeras evidencias de restos óseos humanos en el curso inferior del río Colorado durante el Holoceno medio: sitio Cantera de Rodados Villalonga (pdo. de Patagones, Pcia. de Buenos Aires). Caz Rec Con Sur 6:101–113

Martínez G, Martínez GA, Stoessel L, Alcaráz AP et al (2014) Resultados preliminares del sitio Zoko Andi 1. Aportes para la arqueología del curso inferior del río Colorado (Provinciade Buenos Aires). Rev Mus Antrop 7(1):105–114

Mazzanti D, Quintana C (2001) Cueva Tixi: Cazadores y Recolectores de las sierras de Tandilia Oriental 1. Geología, Paleontología y Zooarqueología. Laboratorio de Arqueología de la Universidad Nacional de Mar del Plata, Mar del Plata

Mengoni Goñalons GL (1999) Cazadores de guanacos de la estepa patagónica. Sociedad Argentina de Antropología, Buenos Aires

Monserrat AL (2010) Evaluación del estado de conservación de dunas costeras: dos escalas de análisis de la costa pampeana. Unpublished PhD thesis, Biblioteca Digital de la Facultad de Ciencias Exactas y Naturales-Universidad de Buenos Aires

Oliva FW (2000) Análisis de las localizaciones de los sitios con representaciones rupestres en el sistema de Ventania, provincia de Buenos Aires. In: Podestá MM, Hoyos M (eds) Arte en las rocas. Arte Rupestre, menhires y piedras de colores en Argentina. Sociedad Argentina de Antropología, Buenos Aires, pp 143–157

Oliva F, Gil A, Roa M (1991a) Recientes investigaciones arqueológicas en el sitio San Martín (BU/PU/5). Partido de Puán, Prov. de Buenos Aires. Shinc 3:135–139

Oliva F, Moirano J, Saghessi M (1991b) Estado de las investigaciones en el sitio Laguna del Puán 1. Bol Cent:127–138

Politis G (1984) Arqueología del Área Interserrana Bonaerense. Unpublished PhD thesis, Facultad de Ciencias Naturales y Museo, Universidad Nacional de La Plata, La Plata

Politis G (2008) The Pampas and Campos of South America. In: Silverman H, William I (eds) Handbook of South American archaeology. Springer, New York, pp 235–260

Politis G, Barros P (2006) La región pampeana como unidad espacial de análisis en la arqueología contemporánea. Fol Hist Nord 16:51–74

Politis G, Bonomo M (2011) Nuevos datos sobre el "hombre fosil" de Ameghino. In: Fernicola J, Prieto A, Lazo D (eds) Vida y obra de Ameghino, Asociación Paleontológica Argentina, vol 12. Publicación Especial, Buenos Aires, pp 101–119

Politis G, Lozano P (1988) Informe preliminar del sitio Costero La Olla (Pdo. De Cnel. Rosales, Pcia. de Buenos Aires). In: Work presented at IX Congreso Nacional de Arqueología Argentina, Buenos Aires

Politis G, Scabuzzo C, Tykot R (2009) An approach to pre-Hispanic diets in the Pampas during the early/middle Holocene. Int J Osteoarchaeol 19:266–280

Politis G, Messineo P, González M et al (2012) Primeros resultados de las investigaciones en el sitio Laguna de Los Pampas (partido de Lincoln, provincia de Buenos Aires). Rel Soc Arg Antr 27(2):7–16

Politis G, Gutiérrez MA, Scabuzzo C (2014) Estado actual de las investigaciones en el Sitio Arqueológico Arroyo Seco 2 (Partido de Tres Arroyos, provincia de Buenos Aires, Argentina). Serie Monográfica del INCUAPA Número 5. FACSO-UNICEN, Olavarría

Quattrocchio M, Borromei EA, Deschamps M et al (2008) Landscape evolution and climate changes in the late Pleistocene-Holocene, southern Pampa (Argentina): evidence from palynology, mammals and sedimentology. Quat Int 181:123–138

Salemme M (1987) Paleoetnozoología del sector bonaerense de Región Pampeana con especial atención a los mamíferos. Unpublished PhD thesis, Facultad de Ciencias Naturales y Museo, Universidad Nacional de La Plata, La Plata

Scabuzzo C (2013) Estudios bioarqueológicos del sitio Paso Mayor, sudoeste de la provincia de Buenos Aires. Rev Mus Antr 6:49–62

Stewart KD, Gifford-González D (1994) An ethnoarchaeological contribution to identifying Hominid fish processing sites. J Archaeol Sci 21:237–341

Stoessel L (2012) Consumo de peces en el área ecotonal árida-semiárida del curso inferior del Río Colorado (provincia de Buenos Aires) durante el Holoceno tardío. Rel Soc Arg Antr 37 (1):159–182

Stoessel L (2015) Tendencias preliminares sobre el consumo de peces durante el Holoceno medio en el área de transición Pampeano-Patagónica Oriental, (provincia de Buenos Aires). Archae. Int J Archaeozo 24:103–117

Vecchi R (2011) Bolas de boleadora en los grupos cazadores- recolectores de la Pampa bonaerense. Unpublished PhD thesis, Facultad de Filosofía y Letras. Universidad de Buenos Aires, Buenos Aires

Vecchi R, Frontini R, Bayón C (2013) Paso Vanoli: una instalación del Holoceno tardío en valles fluviales del sudoeste bonaerense. Rev Mus La Plat Sec Antropo 13(87):77–93

Vecchi R, Frontini R, Bayón C (2014) Ocupaciones en las dunas del litoral atlántico del sudoeste bonaerense: el sitio El Americano II. Paper presented at VII CARPA, Rosario, Universidad Nacional del Rosario, 5–8 Nov 2014

Zohar I, Dayan T, Galili E, Spanier E (2001) Fish processing during early Holocene: a taphonomic case study from Coastal Israel. J Archaeol Sci 28:1041–1053

Shell Mounds of the Southeast Coast of Brazil: Recovering Information on Past Malacological Biodiversity

4

Edson Pereira Silva, Sara Christina Pádua, Rosa Cristina Corrêa Luz Souza, and Michelle Rezende Duarte

4.1 Introduction

Shell mounds (the Brazilian term for it is "sambaquis", a word derived from the Tupi language) are archaeological sites found in almost all coastal areas around the world that, since a little over a century ago, have been recognized as artificial constructions built by prehistoric human populations (Stein 1992). In Brazil, particularly between the states of Espírito Santo (21° S) and Rio Grande do Sul (32° S), there are hundreds of shell mounds that attest the human occupation of the coast between at least 6500 years BP and the start of the Common Era (Lima et al. 2002). Most of these sites are round conical hills of varied size. Its base is usually of a few tens of meters in diameter and often exceeding 2 m high, sometimes surpassing 25 m high.

The sites chosen for shell mound construction are found near embayment, bays and lagoons, on the interface between marine and terrestrial environments, and between salt and fresh water (DeBlasis et al. 2007). The construction of shell mounds in these estuarine environments was not fortuitous, given that these are the environments with the highest biotic productivity on the coast, harboring a high density and diversity of life forms (Lima 2000). The biological remains found in shell mounds include very resistant elements such as mollusk shells, crustacean and sea urchin carapaces, fish, bird and mammal bones (Stein 1992).

In addition to information on prehistoric societies, their food supplies and the use of resources for making ornaments and artifacts, the vestiges found in shell mounds may yield data leading to the examination of other issues. For instance, the fact that

E.P. Silva (✉) • S.C. Pádua • R.C.C.L. Souza • M.R. Duarte
Laboratório de Genética Marinha e Evolução, Departamento de Biologia Marinha, Instituto de Biologia, Universidade Federal Fluminense, Outeiro São João Batista, s/n°, P.O. Box 100644, Niterói, Rio de Janeiro 24001-970, Brazil
e-mail: gbmedson@vm.uff.br; sarapadua@outlook.com; rcclsouza@yahoo.com.br; michellerezendeduarte@yahoo.com.br

© Springer International Publishing AG 2017   47
M. Mondini et al. (eds.), *Zooarchaeology in the Neotropics*,
DOI 10.1007/978-3-319-57328-1_4

these sites contain sets of organisms of the fauna existing at the time of their creation make possible to recover paleoenvironmental aspects related to species biodiversity and biogeography (Lindbladh et al. 2007; Froyd and Willis 2008). Thus, the definition of an environment's pristine state cannot dispense long-term information (Willis and Birks 2006).

Paleoecological studies can also provide data that help the understanding of fundamental issues such as determining the natural expansion of species over time; confirming the status of a species, whether native or exotic; analyzing the rate and patterns of dispersal of invasive species over time; and, finally, assessing the long-term impact of exotic species on native ecosystems (Didham et al. 2005).

In this chapter studies which used archaeozoological remains of marine mollusks from shell mounds of the Central-South coast of Brazil to investigate patterns of biodiversity over space and time are explored. In pursuance, the raw data from the reviewed papers are used in quantitative tests of the resemblance of archaeozoological remains to biocoenoses. The argument defended is that despite any bias associated with the artificial nature of sambaquis, they can still bring some relevant information on Holocene biodiversity and can be useful tools in the efforts of conservation and management.

4.2   A Review

4.2.1   Establishing Baselines

Reference inventories are lists of species that represent the biological diversity against which current changes in biodiversity can be assessed. The study of shell mounds can offer relevant amount of information on species composition, abundances and distribution during the Holocene (Willis et al. 2007). A key objective of these studies is to detect extinctions and declines in population sizes by natural causes or human pressures. Another important inference which can be made from reference inventories of species is the range of variability, characterized in terms of magnitude and frequency, which can be considered as natural on long-term. For example, in a study on Holocene mollusks from Rio de Janeiro State coast in Brazil ($22°24'31''$ S and $23°09'34''$ S) was found a stability of the malacological biodiversity at least in the period between 8000 and 1000 years BP (Souza et al. 2010a). The list of marine species of bivalves and gastropods found in 17 archeological sites of the studied shell mounds showed a total of 124 taxa, of which 65 are bivalves and 59 gastropods, all of them still found today in the region.

Reference inventories of species are also fundamental to the establishment of baselines, which is a concept particularly well-used in ecological restoration where they determine "the ecosystem present before human influence became pronounced on the landscape" (Lindbladh et al. 2007). Although it is acknowledged that environments are dynamic rather than static systems, these baselines are important references to understand biodiversity and environmental changes as well as to guide management and conservation initiatives (Gordillo et al. 2014).

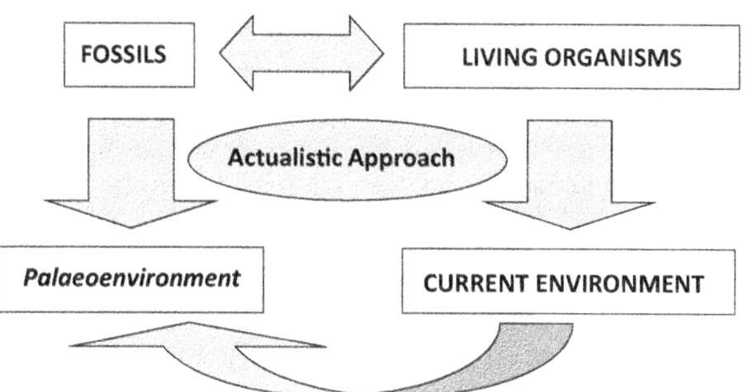

Fig. 4.1 Actualistic approach in which the archaeozoological remains can be used to aid paleoecological interpretations based on the knowledge of the ecological needs of present species

Using an actualistic approach (Fürsich 1995), reference inventories can be used to interpret the ancient conditions and establish environment baselines (Fig. 4.1). Registered organisms from shell mounds are the same or closely related with those inhabiting present-day biological communities and, therefore, environment conditions which govern the establishment and distribution of these communities nowadays should be the same in the past. This approach works very well for relatively young times as those related to Late Holocene. In addition to that, a functional approach, using the morphology of hard parts of the organisms (such as shell from mollusks) can be interpreted according to their function.

Based on these assumptions Souza et al. (2010b), on studying the mollusks fauna from the archaeological site of Sambaqui da Tarioba (22°31′40″ S, 41°56′22″ W), concluded that the Quaternary environmental characteristics were in agreement with current data for the region. Species from the Sambaqui were mainly composed of beach species (59.5%), although mangrove, estuary, and lagoon species were also retrieved. About 85% of the bivalves were unconsolidated substrate species and all were suspensivores. Most gastropods were also unconsolidated substrate species (80%) although, in that case, 60% of the recorded species were carnivorous. These data are in accordance with the present day geomorphological description of the region, which is characterized by sediments associated with fluvial/alluvial and marine deposits, whereas the marine sandy accumulations (marine terraces) and the isolated restingas (sandbanks) are represented by fine to medium-sized, well selected, quartzous sands mixed with clay and organic matter.

The authors concluded that no substantial geomorphological changes seem to have taken place in that region and that the paleoenvironment and past molluskan richness coincide with those seen in the present day.

4.2.2 Bioinvasion

Biological invasions are the arrival, establishment and subsequent spread of non-native species in natural communities. These introductions are usually mediated by human activity and have been increased dramatically as consequence of globalization of trade and shipping transport (Seebens et al. 2013) in the last decades. The introduction of exotic organisms may have irreversible and devastating consequences for natural ecosystems being recognized as one of the greatest threats to biodiversity (McGeoch et al. 2010). This is another key environmental issue that would benefit from the study of shell mounds.

The establishment of baselines based on shell mound studies can help to determine the natural range of species over time, objectively define the status of a species (native *versus* introduced), analyze the rate and pattern of spread of invasive species and assess the long-term impact of alien species on ecosystems (Froyd and Willis 2008). Furthermore, a long temporal record can provide important data for predicting future impacts of invasive species under climate changes (Willis et al. 2007).

In a study of shell mounds of the Southeast coast of Brazil (Souza et al. 2003) a curious fact was noticed: The majority of archaeological vestiges left by hunters and mollusk gatherers, who lived on the shores, show no shells of the species *Perna perna* (Linnaeus, 1758). The species is the most abundant bivalve mollusk on rocky shores of Southeast Brazil, as well as the most commonly consumed organism by the local population. Therefore, it would be expected to be found in the shell mounds. In contrast, *P. perna* is abundantly found in the deepest levels of the shell mounds in the Klasies River-South Africa area (34° 6′ S, 24° 24′ E), in deposits dating from 60,000 to 115,000 BP, the oldest date obtained for this species. One hundred kilometers to the west, near Plettenberg Bay, another archaeological site was found, with 10,000 year-old records of this bivalve (Souza et al. 2004). Added to that, *Pinctada imbricata* (Röding, 1798), a rare species in today's rocky shores, due to its competition for space with *P. perna*, is abundant in the Brazilian sambaquis. In spite of their fragility, *P. imbricata* valves are found in good condition, which seems to indicate that this organism was frequently collected (Lima 1988).

Thus, the absence of consistent fossil records that attest to the existence of *P. perna* in prehistoric times in Brazil, and the existence of recent cases of the invasive behavior of this genus (two biological invasions by populations of the genus *Perna* were reported for the Caribbean and for the Gulf of Mexico; Hicks and Tunnell 1993), lead Souza and collaborators (2003) to raise the hypothesis that the mussel *P. perna* may be an exotic species, possibly introduced in Brazil several years ago. It is also possible that *P. imbricata*, which is rare today on rocky shores but was abundant in prehistoric times, as indicated by its undisputable occurrence in many shell mounds, may have been replaced by the invading *P. perna*.

The authors, then, created a scenario for this bioinvasion event occurred on the Brazilian coast. They argue that during the sixteenth century, Brazil emerged as the greatest destination for African slaves in the Americas, becoming the New World's largest slave importer, a status it maintained during most of the duration of the slave trade to the Americas. Starting in the eighteenth century, ships hailed from Congo, Angola, Mozambique and Tanzania, places where the existence of *P. perna* has been recorded, to the states of Bahia and Rio de Janeiro. Therefore, the authors concluded that *P. perna* arrived at Brazilian coast by this time, having the ships used in the slave trade as a vector. This hypothesis is difficult to be tested; however, it is very ingenious.

4.2.3 Patterns of Biodiversity

To understand the patterns of biodiversity for a given location it is necessary to count not only with an inventory of living organisms, but also an inventory of fossils and sub-fossils of the studied region (Furon 1969). In other words, a comprehensive approach to the biodiversity issue must include the history of a location, placed on an evolutionary perspective. Thus, archaeozoological research is an important tool to recover data on past biodiversity (Tchernov 1992). Despite the fact that shell mounds are artificial accumulations of organisms, the simple presence/absence of species are enough information to create a taxonomic listing which can be used as a helpful tool in defining a historical record of biological diversity (Stahl 2008).

The most extensive study on patterns of biodiversity from the Late Holocene for the Brazilian coast was done for malacological fauna including data of a total of 578 shell mounds located along more than 2000 km along Brazilian coast (Souza et al. 2011, 2012). A total of 154 species were identified, 78 of them being bivalves and 76 gastropods. Species composition in the shell mounds seemed to mirror the present-day pattern of biodiversity. Among bivalves, Veneridae Rafinesque, 1815 is the most diverse family worldwide, with about 50 genera (Mikkelsen and Bieler 2007). In Brazil, 14 genera occur and, of these, nine are represented at Central-South Brazilian shell mounds (64.29%). The family Donacidae Fleming, 1828 is represented by five genera. In Brazil *Donax* Linnaeus 1758 and *Iphigenia* Schumacher, 1817 occur, both represented in the shell mound by the species *Donax hanleyanus* Philippi, 1842 and *Iphigenia brasiliana* (Lamarck, 1818) (Rios 1994). In relation to gastropods, the family Olividae Latreille, 1825 had the greatest diversity, genus *Olivancillaria* d'Orbigny, 1841 being represented by six species (Rios 1994). The family Naticidae Guilding, 1834 presented three genera and six species and Fasciolariidae Gray, 1853 presented three genera and four species.

Souza and collaborators (2012) tried to evaluate changes in biodiversity patterns over time by comparing species and family richness for archaeological sites that are contiguous in space but show an age difference of 2000 years. A conservative procedure was assumed for comparison. From the 578 studied sites only those using ^{14}C radiocarbon dating were taken. Out of them, those with no age overlap were selected. Then a safety margin of about 2000 years difference between sites was chosen. After that, sites with detailed malacological citation were selected. Finally, were used only sites which were very close geographically (10 km maximum distance) to prevent relevant ecological differences among them and include areas in the same exploitation range by the Humans who built up the sambaquis. Following all these procedures, only eight sambaquis remained subject to comparison: Sambaqui Pontinha (22°55′28″ S, 42°31′01″ W), Beirada (22°55′32,04″ S, 42°32′33,46″ W) and Manitiba I (22°55′50,01″ S, 42°34′58,81″ W), Rio de Janeiro. Sambaqui Cosipa 4 (23°52′29″ S, 46°22′21″ W), Cosipa 1 (23°52′29″ S, 46°22′21″ W) and Piaçaguera (23°52′ S, 46°22′ W), São Paulo. Sambaqui Espinheiros II (26°16′59,9″ S, 48°47′35,3″ W) and Morro do Ouro (26°18′51,5″ S, 48°49′37,9″ W), Santa Catarina. The comparison indicates a trend of slight reduction in diversity in more recent sites for all measures used although not for all sites at the same time, suggesting an evolution in biodiversity patterns which could be determined by both cultural (as observed by Amesbury 2007 for mollusks during the prehistoric period in the Mariana Islands) and/or environmental changes (as described by Keegan et al. 2003 for invertebrate taxa at two pre-Columbian sites in Southwestern Jamaica) through time. However, the authors did not put their suggestion under statistical scrutiny.

4.3   Sambaquis as a Proxy of Late Holocene Biodiversity

4.3.1   Bias or Bits?

An ecological proxy (i.e., representative) is a data sampler. Information on past environmental conditions which cannot be directly assessed can be inferred from sources as diverse as farmers' records, naturalists' journey dailies, corals seasonal layers, fluctuations of glaciers and ice caps and growth rings in plant and animals. The main assumption lying under all inferences done by the publications reviewed in the previous section is that sambaquis can be used as proxies for biodiversity inferences. But can they be used as such?

A peculiar and quite evident feature of sambaquis is the fact that they are artificial accumulations of organisms; therefore, the set of organisms found in these sites represent a biased sampling of the natural biological communities. Factors as diverse as culture preferences, technical level, food taboos and the way material were discarded and/or utilized, certainly played a relevant role on the composition of the fauna found in shell mounds. Other questions to be considered when using shell mounds as samples of natural biological communities are differences in the preservation potential of the species in these sites (Prummel and Heinrich 2005). Thus, some investigators believe that shell mound data have selectivity biases that make evaluating the state of historical ecosystems difficult (Baisre 2010).

Biological remains from shell mounds, however, lead to valuable information regarding biodiversity because, despite any bias, these remains represent a sample from the fauna existing at the time of the creation of these archaeological sites (Lindbladh et al. 2007; Froyd and Willis 2008). Furthermore, some shell mound evaluations have been showing detectable changes in species composition over millennia (Dalzell 1998; Maschner et al. 2008), whereas other evaluations show major changes in species composition over a few centuries (Lotze and Milewski 2004; Rosenberg et al. 2005).

Information from shell mounds used in association with modern assemblage inventories have been used to infer changes in life history, diet, and trophic level of fish species assemblages (McClanahan and Omukoto 2011). In the same way, a recent study done on mollusks from the mount of excavation discards from "Sambaqui da Tarioba" (Rio de Janeiro, Brazil) revealed that the taxonomic diversity estimated was representative of the present day mollusk diversity in the coast of Rio de Janeiro state. This result was obtained by means of taxonomic distinctness (AvTD) and variation in taxonomic distinctness (VarTD) tests (Faria et al. 2014).

To test if archaeozoological remains found in sambaquis resemble biocoenoses, the raw data from the reviewed works were compared with censuses of living species of mollusks from the whole spectrum of habitats, both on a local and a regional basis.

4.3.2 Data Sources and Quantitative Analysis

Souza and her collaborators (2003, 2010a, b, 2011, 2012) produced their inventories based on four sources. First, archaeozoological remains integrating collections from nine Brazilian institutions (IPHAN-21ªSR, Museu do Sambaqui da Tarioba-RJ, Museu Nacional/UFRJ, Instituto de Arqueologia Brasileira-RJ, Museu Arqueológico de Itaipu-RJ, Museu de Arqueologia e Etnologia-UFPR, Museu Arqueológico de Sambaqui de Joinville-SC, Museu Universitário Oswaldo Rodrigues Cabral-UFSC and Museu do Homem do Sambaqui "Padre João Alfredo Rohr, SJ"-SC) were analyzed. In addition, a survey of all publications related to shell mounds from the Central-South Brazilian coast was carried out in the libraries of the institutions where the collections were examined. This survey aimed at finding malacological mentions for additional sites within the studied area and lacking collections of archaeological material. Excavations of four sites were also carried out. Besides excavations, *in situ* observations were done at four other sites which have exposed layers of mollusk shells. Details on all procedures followed to obtain data for each source can be found in the original publications.

Inventories of living species were created based on Rios (1994), which is the most complete census of the Brazilian malacological fauna, using the following criteria: (a) distribution of the species in the area comprised between the same latitudes of Rio de Janeiro coast and Central-South Brazil (it was not possible to produce a inventory of mollusks specific for the latitudes where the Sambaqui da Tarioba is located); (b) shell larger than 5 mm; (c) not being exclusive to oceanic islands; (d) benthic; and, (e) occurring until 20 m in depth.

In the end it was possible to obtain six inventories: three related to sambaquis (Sambaqui da Tarioba, Souza et al. 2010b; Rio de Janeiro coast, Souza et al. 2003, 2010a; and, Central-South Brazil, Souza et al. 2011, 2012); two for the living mollusk fauna (Rio de Janeiro coast and Central-South Brazil, both available under request to the authors) and one obtained from data on 12 malacological surveys on the Brazilian coast (given in Souza et al. 2010b, p. 368, Table 4.2). Figure 4.2 shows the geographical location of the three studied scales.

Quantitative analyses were done in three different ways. First, contingency tables, which test if the proportions of taxa are statistically different between sambaqui and living species inventories, were used. This is a test for differences in taxonomic diversity (Sokal and Rohlf 1997). Secondly, compositional quantitative fidelity was estimated for all taxa. Quantitative fidelity deals with comparisons between living and death assemblages (here, sambaquis), in order to quantitatively assess the preservation of the signal of species composition from the former in the latter. It is focused on the

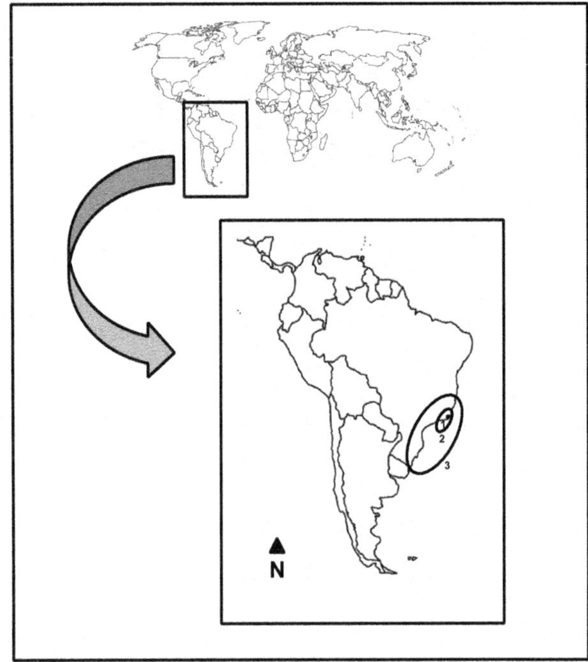

Fig. 4.2 Geographical indication of the three studied scales is shown in the map: 1-Location of the Sambaqui da Tarioba, in the municipality of Rio das Ostras, state of Rio de Janeiro, Brazil (22° 31′40″ S and 41° 56′ 22″ W); 2-State of Rio de Janeiro, Brazil (22° 24′ S till 23° 09′S); 3-Central-South Brazilian coast (20° 09′ S till 30° 32′S)

two indices, the so-called F1 (Eq. 4.1), the percentage of species found alive that were also found dead (a measure of the preservation of the signal of species composition from biocoenose); and F2 (Eq. 4.2), the percentage of species found dead that were also found alive (a measure of the spatial signal of skeletal remains to their original biocoenose); where NS = number of species shared by the two assemblages; ND = number of species found only in death assemblages; and NL = number of species found only in living assemblages (Ritter and Erthal 2013).

$$F1 = \frac{NS \times 100}{NS + NL}$$

(4.1)

$$F2 = \frac{NS \times 100}{NS + ND}$$

(4.2)

Finally, species richness at micro (578 sambaquis), meso (32 localities, presented in Table 4.1, clustered from the 578 sambaquis based on spatial and ecological similarities) and macroregional (6 Brazilian Federative Unions which represent a random cluster) scales were ranked and compared with the 12 malacological surveys from the Brazilian coast given in Souza et al. (2010b). Comparisons at the microregional scale were direct. However, due to lack of surveys for the meso and macro scales, square and cube roots transformations were used in order to simulate the gradual increment of species for each level. For these analyses, a standardized variance of species richness was used following the Eq. (4.3); where $\sigma^2 =$ variance of species richness for the level; and, $\mu =$ the average of species richness for the level same level (Wright 1978). Ranked variances from sambaquis and living assemblages were tested for significant differences by means of U Mann-Whitney Test (for two by two comparisons) and Kruskal-Wallis (for overall comparisons).

$$R = \frac{\sigma^2}{\mu(1-\mu)} \tag{4.3}$$

All the analyses were undertaken using the software packages *Excel for Windows 2007* and *Past 2.08* (Hammer et al. 2001).

Table 4.1 Sambaquis clustered in 32 localities based on spatial and ecological similarities (names of Sambaquis are given in Souza et al. 2011, pp. 28–43)

Locality	Sambaquis
1	1–5
2	6
3	7–17
4	18
5	19–23
6	24–33
7	34–52
8	53, 54
9	55
10	56
11	57, 58
12	59–62
13	63–74
14	75, 76
15	77–79
16	80–83
17	84–98
18	99
19	100–140
20	141, 142
21	143
22	144–156
23	157–203
24	204–208
25	209
26	210–223
27	224
28	225–227
29	228, 229
30	230–427
31	428–554
32	555–578

4.3.3 Artificial Accumulations But Not So Much

The documentary quality of shell mound archaeozoological remains is an important issue for biologists, who increasingly look at accumulations of bones, shells, and plant material as possible ways to extend the time-frame of observation on patterns of biodiversity. Regarding taxonomic diversity, shell mounds very much resemble biocoenoses from the same area for all Linnaean taxa scored, but in upper and lower levels for the whole of Central-South Brazilian coast (Table 4.2). This result is in accordance with previous tests (Faria et al. 2014), which showed that the malacological inventory from Sambaqui da Tarioba was able to recover biodiversity patterns from the modern fauna inhabiting the Rio de Janeiro coast. Although covering two more scales (meso and macroregional) in comparison with the previous work, the test undergone here did not use randomizations and, consequently, it is not as powerful as that one.

The fidelity of shell mound data (the extent to which archaeozoological remains reflect their source biota) is the key measure of their quality as proxies for biodiversity studies. However, a search on this issue (shell mounds and fidelity) showed no results in a search done in *Web of Knowledge* and *Google Scholar* (November 25, 2015). Concerning death assemblages, studies are extensive (Cummins 1994; and references therein) and despite fidelity appears to vary widely, meta-analysis suggests that much of the variance arises from differences in the quality of live data (especially single census *versus* replicates over time, Kidwell and Flessa 1995). Thus, the general conclusion has been that death assemblages are good testimonies of the past biodiversity (Kidwell and Flessa 1995; Brown et al. 2005; Kidwell 2008). Applying the same fidelity index as those used for death assemblages, sambaquis performed as well as death assemblages for F1 measure and even better for F2 (Table 4.3). Molluskan death assemblages typically show F1 varying from 45 to 100% (average = 70%) as for F2 measures ranges from 27 to 100% (average = 49%) (Cummins 1994).

Table 4.2 Taxonomic diversity found in Sambaqui da Tarioba (ST), Rio de Janeiro (RJC) and Central-South coast (CSC) of Brazil in relation to living fauna of mollusks for the same areas (RJC-LF and CSC-LF)

	Sambaquis			Living fauna		χ^2		
Taxa	ST	RJC	CSC	RJC-LF	CSC-LF	ST × RJC-LF	RJC × RJC-LF	CSC × CSC-LF
Species	47	117	148	404	651	0.10	0.12	5.8×10^{-10}
Genera	39	92	99	223	334			
Familiae	28	54	59	94	137	0.63	0.45	0.09
Ordines	7	15	17	18	19			
Bivalves	27	65	75	188	254	0.17	0.13	0.02
Gastropodes	20	52	73	216	397			

The probability of taxonomic diversity be different between inventories (sambaquis *versus* living fauna) is given based on a χ^2 test

Table 4.3 Compositional quantitative fidelity of the sambaquis

Taxa	Sambaqui da Tarioba		Rio de Janeiro coast		Central-South coast	
	F1	F2	F1	F2	F1	F2
Species	53.1	99.8	58.0	95.7	56.2	96.8
Genera	54.9	100.0	63.1	96.7	58.6	97.2
Familiae	56.3	100.0	69.1	98.0	62.3	96.6
Ordines	64.3	100.0	80.0	90.9	77.4	82.8
Bivalves	53.4	99.5	60.1	97.0	58.2	96.0
Gastropodes	52.8	100.0	56.3	94.6	54.9	97.4

F1 = % species found in the living fauna which are also found at the sambaquis, F2 = % species found in sambaquis which are also found at the living fauna

Table 4.4 Standardized variance of species richness for micro (sambaquis/surveys), meso (localities) and macro (Federation Units) scales

	Sambaquis (S)	Living fauna (LF)	S + LF
Sambaquis/surveys	0.080321388	0.273780199	0.091535582
Localities	0.102794635	0.175622131	0.008843641
Federation Unit	0.009004663	0.125161168	0.008527471

Sewall Wright (1889–1988), a leading population geneticist from the time of evolutionary synthesis, developed a way of estimating the genetic population structure among populations. The method used was rather simple although very ingenious. By using a standardized variance of gene frequencies he could infer how isolated populations were and, therefore, differentiate them independently by the play of the evolutionary forces (Wright 1978). The same rationalization can be used in testing the effects of different sampling (or localities or random associations) over species richness (which is related to sampling effort). One would expect that random sampling (for example, a scientific survey of molluskans) would produce lower variances than selective sampling (for example, artificial accumulation in a sambaqui). Furthermore, the different matrices of variances (random sampling *versus* artificial accumulation) can be compared for significant statistical differences (Table 4.4). This test was performed and it was not possible to find significant statistical differences for any of the comparisons [Sambaquis *versus* Living Fauna U-Mann Whitney = 0.1; Sambaquis *versus* (Sambaquis + Living Fauna) U-Mann Whitney = 0.4; Living Fauna *versus* (Sambaquis + Living Fauna) U-Mann Whitney = 0.1; Sambaquis *versus* Living Fauna *versus* (Sambaquis + Living Fauna) Kruskal Wallis = 0.0509)]. Statistical significance considered was 0.05.

Modern ecological studies investigating ecosystem responses to environmental changes in marine systems typically use a perspective restricted to annual or decadal timescales, because quantitative biological monitoring data are not available for time intervals preceding the onset of extensive anthropogenic ecosystem modification (Jackson et al. 2001). Nonetheless, the relative recent past can be used to identify environmental changes beyond timescales of direct human experience if one can use sambaquis as proxies. The results presented here, although modest, seem promising in this sense, namely that sambaquis are artificial accumulations but not so much.

4.4 Conclusion

Brazilian zooarchaeology classical studies have attempted to explain human settlement and cultural trajectories in different geographical regions. In this chapter, some efforts to use archaeozoological remains in order to understand the evolution of the Brazilian Southeast coastal marine biodiversity were reviewed. Furthermore, quantitative analysis using the raw data of these researches and comparisons with censuses of living species were undertaken in order to test the assumption embraced by them. Results support the case that sambaquis can be useful proxies for Late Holocene biodiversity, at least for mollusks.

Acknowledgements The authors would like to thank CAPES (Coordenação de Aperfeiçoamento de Pessoal de Nível Superior) for financial support (PNPD-Programa Nacional de Pós Doutorado) and scholarships for RCCL Souza (Post-doctorate) and MR Duarte (PhD). SC Pádua is financially supported by a CNPq (Conselho Nacional de Desenvolvimento Científico e Tecnológico) scholarship for Science Juniors (PIBIC-Programa Institucional de Bolsas de Iniciação Científica).

References

Amesbury JR (2007) Mollusk collecting and environmental change during the prehistoric period in the Mariana Islands. Coral Reefs 26:947–958

Baisre J (2010) Setting a baseline for Caribbean fisheries. J Island Coast Archaeol 5:120–147

Brown ME, Kowalewski M, Neves RJ, Cherry DS, Schreiber ME (2005) Freshwater mussel shells as environmental chronicles: geochemical and taphonomic signatures of mercury-related extirpations in the North Fork Holston River, Virginia. Environ Sci Technol 39:1455–1462

Cummins RH (1994) Taphonomic processes in modern freshwater molluscan death assemblages: implications for the freshwater fossil record. Palaeogeogr Palaeoclimatol Palaeoecol 108:55–73

Dalzell P (1998) The role of archaeological and cultural-historical records in long-range coastal fisheries resources management strategies and policies in the Pacific Island. Ocean Coast Manag 40:237–252

DeBlasis P, Kneip A, Scheel-Ybert R, Giannini PC, Gaspar MD (2007) Sambaquis e paisagem: Dinâmica natural e arqueologia regional no litoral do sul do Brasil. Arqueol Suramericana 3:29–61

Didham RK, Tylianakis JM, Hutchison MA, Ewers RM, Gemmel NJ (2005) Are invasive species the drivers of ecological change? Trends Ecol Evol 20:470–474

Faria RGS, Silva EP, Souza RCCL (2014) Biodiversity of Marine Molluscs from Sambaqui da Tarioba, Rio das Ostras, Rio de Janeiro (Brazil). Rev Chilena de Antropol 29(1):49–54

Froyd CA, Willis KJ (2008) Emerging issues in biodiversity & conservation management: the need for a palaeoecological perspective. Quat Sci Rev 27:1723–1732

Furon R (1969) La distribución de los seres. Editorial Labor, Barcelona

Fürsich FT (1995) Approaches to palaeoenvironmental reconstructions. Geobios 18:183–195

Gordillo S, Bayer SB, Boretto B, Charó M (2014) Mollusk shells as bio-geo-archives. Evaluating environmental changes during the quaternary. Springer, Cham

Hammer Ø, Harper DAT, Ryan PD (2001) PAST: Paleontological Statistics Software Package for education and data analysis. Palaeontol Electron 4(1):1–9

Hicks DW, Tunnell JW (1993) Invasion of the south Texas coast by the edible brown mussel *Perna perna* (Linnaeus, 1758). Veliger 36:92–97

Jackson JBC, Kirby MX, Berger WH, Bjorndal KA, Botsford LW, Bourque BJ, Bradbury RH, Cooke R, Erlandson J, Estes JA, Hughes TP, Kidwell S, Lange CB, Lenihan HS, Pandolfi JM, Peterson CH, Steneck RS, Tegner MJ, Andwarner RR (2001) Historical overfishing and the recent collapse of coastal ecosystems. Science 293:629–637

Keegan WF, Portell RW, Slapcinsky J (2003) Changes in invertebrate taxa at two pre-Columbian sites in southwestern Jamaica, AD 800-1500. J Archaeol Sci 30:1607–1617

Kidwell SM (2008) Ecological fidelity of open marine molluscan death assemblages: effects of post-mortem transportation, shelf health, and taphonomic inertia. Lethaia 41:199–217

Kidwell SM, Flessa KW (1995) The quality of the fossil record: populations, species, and communities. Annu Rev Ecol Evol Syst 26:269–299

Lima TA (1988) Pérolas milenares. Ciência Hoje 7(42):66–67

Lima TA (2000) Em busca dos frutos do mar: os pescadores-coletores do litoral centro-sul do Brasil. Revista USP 44:270–327

Lima TA, Macário KD, Anjos RM, Gomes PRS, Coimbra MM, Elmore E (2002) The antiquity of the prehistoric settlement of the central-south Brazilian coast. Radiocarbon 44(3):733–738

Lindbladh M, Brunet J, Hannon G, Niklasson M, Eliasson P, Eriksson G, Ekstrand A (2007) Forest history as a basis for ecosystem restoration: a multidisciplinary case study in a south Swedish temperate landscape. Restor Ecol 15:284–295

Lotze HK, Milewski I (2004) Two centuries of multiple human impacts and successive changes in a North Atlantic food web. Ecol Appl 14:1428–1447

Maschner HDG, Betts MW, Reedy-Maschner KL, Trites AW (2008) A 4500-year time series of Pacific cod (*Gadus macrocephals*) size and abundance: archaeology, oceanic regime shifts, and sustainable fisheries. Fish Bull 104:386–394

McClanahan TR, Omukoto JO (2011) Comparison of modern and historical fish catches (AD 750–1400) to inform goals for marine protected areas and sustainable fisheries. Conserv Biol 25(5):945–955

McGeoch MA, Butchart SHM, Spear D, Marais E, Kleynhans EJ, Symes A (2010) Global indicators of biological invasion: species numbers, biodiversity impact and policy responses. Divers Distrib 16:95–108

Mikkelsen PM, Bieler R (2007) Seashells of Southern Florida: living marine bivalves of the Florida keys and adjacent regions. Princeton University Press, Princeton

Prummel W, Heinrich D (2005) Archaeological evidence of former occurrence and changes in fishes, amphibians, birds, mammals and molluscs in the Wadden Sea area. Mar Res 59 (1):55–70

Rios EC (1994) Seashells of Brazil. FURG, Rio Grande

Ritter MN, Erthal F (2013) Fidelity bias in mollusk assemblages from coastal lagoons of Southern Brazil. Rev Bras Paleontol 16(2):225–236

Rosenberg AA, Bolster WJ, Alexander KE, Leavenworth WB, Cooper AB, McKenzie MG (2005) The history of ocean resources: modeling cod biomass using historical records. Front Ecol Environ 3:84–90

Seebens H, Gastner MT, Blasius B (2013) The risk of marine bioinvasion caused by global shipping. Ecol Lett 16:782–790

Sokal RR, Rohlf FJ (1997) Biometry—the principles and practice of statistics in biological research. W.H. Freeman, New York

Souza RCCL, Fernandes FC, Silva EP (2003) A study on the occurrence of the brown mussel *Perna perna* on the sambaquis of the Brazilian coast. Revista do Museu de Arqueologia e Etnologia 13:3–24

Souza RCCL, Fernandes FC, Silva EP (2004) Distribuição atual do mexilhão *Perna perna* no mundo: um caso recente de bioinvasão. In: Silva JSV, Souza RCCL (orgs) Água de Lastro e Bioinvasão. Editora Interciência, Rio de Janeiro, pp 157–172

Souza RCCL, Lima TA, Silva EP (2010a) Holocene molluscs from Rio de Janeiro state coast, Brazil. Check List 6(2):301–308

Souza RCCL, Trindade DC, Decco J, Lima TA, Silva EP (2010b) Archaeozoology of marine mollusks from Sambaqui da Tarioba, Rio das Ostras, Rio de Janeiro, Brazil. Fortschr Zool 27 (3):363–371

Souza RCCL, Lima TA, Silva EP (2011) Conchas Marinhas de Sambaquis do Brasil. Technical Books Editora, Rio de Janeiro

Souza RCCL, Lima TA, Silva EP (2012) Remarks on the biodiversity of marine molluscs from Late Holocene Brazilian shell mounds. In: Lefèvre C (ed) Proceedings of the general session of the 11th International Council for Archaeozoology conference (Paris, 23–28 August 2010), BAR international series 2354. Archaeopress, Oxford, pp 245–256

Stahl PW (2008) The contributions of zooarchaeology to historical ecology in the neotropics. Quat Int 180:5–16

Stein JK (1992) The analysis of shell middens. In: Stein JK (ed) Deciphering a shell midden. Academic Press, San Diego, pp 1–24

Tchernov E (1992) Evolution of complexities, exploitation of the biosphere and zooarchaeology. Archaeozoologia 5(1):9–42

Willis KJ, Birks HJB (2006) What is natural? The need for a long-term perspective in biodiversity and conservation. Science 314:1261–1265

Willis KJ, Araújo MB, Bennett KD, Figueroa-Rangel B, Froyd CA, Meyers N (2007) How can a knowledge of the past help to conserve the future? Biodiversity conservation and the relevance of long-term ecological studies. Philos Trans R Soc B 362:175–186

Wright S (1978) Evolution and the genetics of populations. Volume 4. Variability within and among natural populations. University of Chicago Press, Chicago

Paula D. Escosteguy and Mónica C. Salemme

5.1 Introduction

Cañada Honda is an archaeological locality composed of seven sites on both margins of the Cañada Honda creek and the Río Areco (Baradero, Northeastern Buenos Aires province; Fig. 5.1). It was discovered in 1948 and excavated during the following years (Bonaparte 1951). The materials recovered are available at the Museo Municipal 'Carlos Ameghino', in the city of Mercedes, Buenos Aires Province. Most archaeological elements come from Site 1, but the specific source of some of the others—within the locality—is unknown. In this sense, a collection of materials from this archaeological locality did not achieve all modern parameters of scientific fieldwork (*sensu* Migale and Bonaparte 2008). Nevertheless, the faunal collection is studied as a whole, taking into account their relevance and contemporaneity, being all materials dated to the Late Holocene.

This locality is surrounded by the wetlands riverine environment of the lower río Paraná (Fig. 5.2). This area between the Paraná Delta and the Pampa Ondulada ("Undulate Pampa") is known as 'Bajíos Ribereños' (*sensu* Bonfils 1962; Acosta and Mucciolo 2014), where a high number of archaeological sites have been reported (see Acosta 2005; Loponte 2007). Usually, they are located in fluvial banks that raise between a few centimeters to 1 m above the surrounding landscape, which could have protected the hunter-gatherer groups from successive floods, a frequent feature on these fluvial environments. Archaeological findings came from

P.D. Escosteguy (✉)
Instituto de Arqueología, FFyL, Universidad de Buenos Aires-CONICET, 25 de Mayo 217, 3er. piso, 1002 Ciudad Autónoma de Buenos Aires, Argentina
e-mail: paueguy@hotmail.com

M.C. Salemme
Centro Austral de Investigaciones Científicas—CONICET and Universidad Nacional de Tierra del Fuego, B. Houssay 200, 9410 Ushuaia, Argentina
e-mail: msalemme@cadic-conicet.gob.ar

© Springer International Publishing AG 2017
M. Mondini et al. (eds.), *Zooarchaeology in the Neotropics*,
DOI 10.1007/978-3-319-57328-1_5

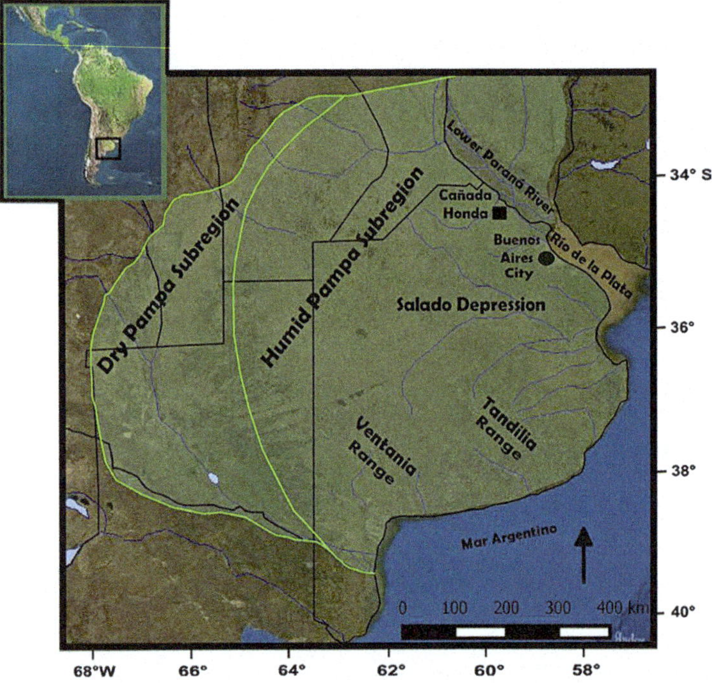

Fig. 5.1 Location of Cañada Honda Archaeological Locality

Fig. 5.2 Landscape surrounding Cañada Honda Archaeological Locality (photograph by Sonia Lanzelotti 2008)

Layer A of the present soil level in all of the sites (Loponte 2007; Arrizurieta et al. 2010).

The archaeological deposits from Cañada Honda contained a variety of materials such as ceramics, lithics, and bone artifacts, in addition to faunal remains and, at least four human skeletons with gravegoods (Bonaparte 1951; Pérez Jimeno 2004; Migale and Bonaparte 2008; Lanzelotti and Bonaparte 2009; Lanzelotti and Acuña 2010; Lanzelotti et al. 2011). Two bones of *Lama guanicoe* from Site 1 were radiocarbon dated (see Lanzelotti et al. 2011) to 2030 ± 100 years C^{14} B.P. (LP-2368) and 2130 ± 60 years C^{14} B.P. (LP-2422).

The numerous faunal remains were partially analyzed in previous works (Salemme 1987, 1990), including studies on rodents and ungulates (Escosteguy 2011, 2012; Escosteguy and Salemme 2012; Escosteguy et al. 2012; Salemme et al. 2012). The aim of this contribution is to review the complete faunal collection focusing on the economic strategies followed by the hunter-gatherer groups who inhabited Cañada Honda during the Late Holocene.

This period reveals particular characteristics in the Pampas; multiple cultural developments have influenced the interactions between humans and the environment. Some innovations as arrows and spears, bone technology, pottery vessels, could have offered new possibilities in the exploitation of faunal and plant resources in a way that led people to maximize benefits (Martínez and Gutiérrez 2004; Acosta 2005; González 2005; Loponte 2007; Quintana and Mazzanti 2014; Stoessel and Martínez 2014). Economic strategies of diversification and specialization were closely related to other cultural developments in each different Pampean areas (Quintana and Mazzanti 2014; Stoessel and Martínez 2014; and references cited there). By this time, an increase of intra- and inter-regional exchange is well documented; it is revealed through the record of species from other environments or lithic raw materials from distant places as the río Uruguay and the Tandilia and Ventania ranges, in Buenos Aires province (González 2005; Quintana and Mazzanti 2014; among others). In addition, demographic increase took place and, consequently, longer or recurrent occupation of the same sites (Quintana et al. 2002; González 2005).

5.2   Methodology

The methodological approach used included taxonomic and anatomic identification. Taxonomic abundance was estimated using NSP (Number of Specimens), NISP (Number of Identified Specimens), %NISP (% Number of Identified Specimens) and MNI (Minimum Number of Individuals) (Grayson 1984; Lyman 1994, 2008; Mengoni Goñalons 1999). In this paper, two size categories for animals were considered: 'small fauna' and 'large fauna'. The former includes vertebrates weighing less than 15 kg whereas the latter involves the animals over that weight limit.

Naked-eye analysis of the marks visible on bone surfaces was performed using hand lens and/or low-power microscope (between 7.5X and 35X). For natural marks such as weathering, trampling and abrasion, proposals of different authors were considered: Behrensmeyer (1978), Andrews (1990) and Gutiérrez (2004). In the case of biological agents, rodent gnawing, roots, carnivore gnawing and corrosion by digestive acids were recorded as well (Andrews 1990; Mengoni Goñalons 1999; Gómez 2000; Quintana 2007; Álvarez et al. 2011).

Natural modifications could have obliterated cultural marks on bone surfaces. In some cases, manganese oxide had dyed the bone so extensively that it was difficult to distinguish the action of fire, for instance. Root or dermestid pits could have intersected and covered cutmarks as well. Carnivores, like foxes living in the Pampas, may be modified bone assemblages, especially those of small animals. These mammals could have conditioned the anatomical profiles and even destroyed bone surfaces (Álvarez et al. 2011).

Regarding cutmarks, criteria by Binford (1981), Fisher (1995) and Mengoni Goñalons (1999) were followed, whereas fractures were analyzed under Miotti and Salemme (1988), Pérez Ripoll (2005/2006) and Mengoni Goñalons (1999) proposals. The primary fracture type (helical, longitudinal and transverse) was considered in as much as other features as negative flakes, notches, impact points and fracture surface. For heat alterations, a scale of color was used to define burnt (reddish-brown), carbonized (black), calcined (bluish-gray/white) and unburnt (whitish-yellowish) specimens (Mengoni Goñalons 1999); thermal alterations were considered evidence of fauna management.

5.3   Zooarchaeological and Taphonomical Analysis

5.3.1   Identification

Vertebrates of several sizes were identified in the assemblage from the Cañada Honda locality (Table 5.1). Among birds, *Rhea americana* (American ostrich-like, ñandú) was outstanding for its size. Large mammals were abundant and diverse; there were specimens of Artiodactyla as *Lama guanicoe* (guanaco), *Blastoceros dichotomus* (marsh deer) and *Ozotoceros bezoarticus* (Pampean deer), as well as some others broadly identified as Cervidae. The largest South American rodent, capybara (*Hydrochaeris hydrochaeris*), was also present in the record. A few bones identified as *Bos taurus* (European cow) were not included in the analysis as they were regarded intrusive in the deposit. These large vertebrates represented 35% of the assemblage, whereas small fauna amounted for 65% of the total (Table 5.1).

Regarding the latter, fish bones from Siluriformes were abundant and probably most of them belonged to the genus *Pterodoras*; others remains were classified as undeterminable fish. A small number of reptilian bones corresponding to a lizard were identified as *Tupinambis merianae*.

Bird bones correspond to species from both aquatic and terrestrial habits. Bird species of terrestrial habits were represented by the family Tinamidae, with a considerable amount of bones of red-winged tinamou (*Rhynchotus rufescens*). A

Table 5.1 NISP, %NISP and MNI of the taxa recorded in Cañada Honda

Size	Class	Order	Family	Taxon	NISP	%NISP	MNI
Large fauna	AVES	Rheiformes	Rheidae	*Rhea americana*	9	0.31	1
		Rodentia	Caviidae	*Hydrochaeris hydrochaeris*	11	0.38	2
	MAMMALIA	Artiodactyla	Cervidae	*Ozotoceros bezoarticus*	118	4.11	10
				Blastoceros dichotomus	39	1.36	4
				Cervidae undet.	24	0.83	–
			Camelidae	*Lama guanicoe*	73	2.54	4
			Undet.	Artiodactyla	27	0.94	–
		Undet.	Undet. large mammal		691	24.08	–
Subtotal					**992**	**34.55**	**21**
Small fauna	PISCES	Siluriformes	Doradidae	cf. *Pterodoras*	33	1.15	–
		Undet.		PISCES undet.	124	4.32	–
	Reptilia	Squamata	Teiidae	*Tupinambis merianae*	11	0.38	1
		Undet.		Reptilia undet.	5	0.17	–
	AVES	Gruiformes	Rallidae	*Fulica leucoptera*	2	0.07	1
				Fulica armillata	10	0.35	2
				Fulica sp.	3	0.1	1
		Pelecaniformes	Phalacrocoracidae	*Phalacrocorax olivaceus*	1	0.03	1
			Ardeidae	Ardeidae undet.	1	0.03	–
		Anseriformes	Anatidae	*Anas sibilatrix*	6	0.2	2
				Anas versicolor	4	0.14	1
				Anas platalea	3	0.1	1
				Anas georgica	4	0.14	1
				Dendrocygna sp.	4	0.14	1
				Dendrocygna bicolor	2	0.07	1
				Cygnus melancoryphus	1	0.03	1

(continued)

Table 5.1 (continued)

Size	Class	Order	Family	Taxon	NISP	%NISP	MNI
				Chloephaga picta	1	0.03	1
				Anatidae undet.	1	0.03	–
			Anhimidae	*Chauna torquata*	4	0.14	1
		Charadriiformes	Laridae	Laridae undet.	1	0.03	–
		Tinamiformes	Tinamidae	*Nothura maculosa*	7	0.24	3
				Rhynchotus rufescens	27	0.94	5
				Tinamidae undet.	3	0.1	–
		Undet.		AVES undet.	17	0.6	–
	MAMMALIA	Cingulata	Dasypodidae	*Chaetophractus villosus*	8	0.28	2
				Dasypodidae undet.	2[a]	0.07	–
		Rodentia	Undet.	Rodentia undet.	69	2.4	–
			Chinchillidae	*Lagostomus maximus*	84	2.92	12
			Echymidae	*Myocastor coypus*	764	26.63	52
			Octodontinae	*Ctenomys* sp.	2	0.07	1
			Caviidae	*Cavia aperea*	593	20.7	145
				Caviidae undet.	3	0.1	–
		Carnivora	Mustelidae	*Galictis cuja*	3	0.1	1
				Mustelidae undet.	3	0.1	–
			Canidae	*Lycalopex gymnocercus*	4	0.14	1
				Canidae undet.	1	0.03	–
			Felidae	*Felis geoffroyi*	1	0.03	–
				Felidae undet.	1	0.03	–
		Didelphimorfia	Didelphidae	Didelphidae undet.	9	0.31	–
		Undet.		Undet. medium mammal	48	1.67	–
				Undet. small mammal	9	0.31	–

Subtotal	1875	65.35	237
Total	2867	99.9	258
Undet.	45		
NSP	2912		

[a]Plates

few bones of spotted nothura (*Nothura maculosa*) and other specimens assigned to this family level completed this category.

Coots like *Fulica leucoptera* (white-wing coot), *Fulica armillata* (red-gartered coot) and *Fulica* sp.; ducks like *Anas platalea* (red shoveller), *Anas georgica* (yellow-billed pintail), *Anas sibilatrix* (southern wigeon), *Anas versicolor* (silver teal), *Chloephaga picta* (upland goose), *Dendrocygna* sp. and *Dendrocygna bicolor* (whistling duck), and *Cygnus melancoryphus* (black-necked swan) were detected. Other Anseriformes included *Chauna torquata* (southern screamer). Birds from the order Pelecaniformes were also recorded, like *Phalacrocorax olivaceus* (neothropic cormorant) and herns from the family Ardeidae. Scarce bones assigned to Laridae were present as well.

Within the small fauna, mammals were the most diverse class, and rodents outnumbered the rest of the animals. Small rodents (less than 3 kg) like *Cavia aperea* (guinea pig) and *Ctenomys* sp. (tuco-tuco), as much as medium sized rodents like *Myocastor coypus* (coypu) and *Lagostomus maximus* (plains viscacha) were also frequent. Armadillos such as *Chaetophractus villosus* (large hairy armadillo) and other dasypodids were present in lower numbers. Among carnivores, there were canids (*Lycalopex gymnocercus*, pampas fox), felids (*Leopardus geoffroyi*, Geoffroy's cat), mustelids (*Galictis cuja*, lesser grison) and a few bones of Mustelidae and Didelphidae were also recorded.

5.3.2 Anthropic Marks on Bones

The analysis of marks and damage on bone surfaces aided the interpretation of human exploitation of both wetlands and inland species recorded. Cutmarks, scraping marks and fractures of anthropogenic origin were identified on different taxa (Table 5.2).

5.3.2.1 Large Fauna

Evidence of anthropic action on *R. americana* specimens was found on a phalanx and some tarsometatarsals that showed several helical and one straight furrow fracture around the perimeter of the bone. Furthermore, one fracture edge was smooth and polished.

From the bones assigned to Artiodactyla, only one humerus yielded cutmarks running parallel to the transversal and straight fracture, maybe as a result of a furrow along the bone perimeter. Among the specimens identified as Cervidae, a cutmark on a rib and two anthropic fractures were evident (a helical one with beveled surface fractures and a straight mark defining probable furrowing around the bone). Furthermore, only one case showed flake scars.

A short fragment of *B. dichotomus* antler presented cutmarks, together with a metacarpal and a distal metapodial; the latter also yielded a sawed furrow running along part of the bone perimeter (Fig. 5.3a). The rest of the fractures recorded were helical with beveled surface fractures, as well as longitudinal parallel ones. Regarding *L. guanicoe* bones, a fracture on green bone and flake scars in a phalanx were

Table 5.2 Taxa with evidence of exploitation: anthropic marks and anthropic fractures, represented in NISP

Size	Taxa	Anthropic marks	Fractures[a]
Large fauna	*R. americana*	0	4
	H. hydrochaeris	1	0
	O. bezoarticus	9	20
	B. dichotomus	3	3
	Cervidae undet.	1	3
	L. guanicoe	8	8
	Artiodactyla	1	1
	Large size mammal	25	10
Small fauna	*P. olivaceus*	1	0
	A. georgica	1	0
	D. bicolor	1	0
	Anatidae undet.	1	0
	R. rufescens	1	0
	Tinamidae undet.	1	0
	AVES undet.	1	1
	L. maximus	2	7
	M. coypus	37	76
	Rodentia undet.	2	0
	L. gymnocercus	1	1
	Small size mammal	0	3

[a]It includes fractures and associated features such as negative flakes

observed. Likewise, some bones of the smallest deer (*O. bezoarticus*) presented cutmarks, as those registered on the mandible, lumbar vertebrae (Fig. 5.3b), humerus, femur and metatarsals. Some of these cutmarks were associated to green fractures or flake scars; other fractures could be the result of furrows along the bone perimeter, while a metapodial showed a notch of likely anthropic origin. *L. guanicoe* remains also had cutmarks on metacarpals, metapodials and a rib. Finally, *H. hydrochaeris* showed cutmarks only on some talus faces (Fig. 5.4).

Regarding bones as raw material, Pérez Jimeno (2004) recorded 19 bone tools in the assemblage from Cañada Honda, all of them made on large mammal bones. Two artifacts were crafted on *O. bezoarticus* bones, and other two instruments, on *B. dichotomus* specimens (one of them on an antler fragment). Additionally, some *L. guanicoe* bones were decorated and others used to manufacture different tools. A discoidal artifact and a spoon were performed on large mammal bones, but the species could not be identified (Salemme 1987; Pérez Jimeno 2004).

Some bones of both cervid species and *L. guanicoe* are thermally-altered bones (Fig. 5.5). Despite their low percentages (around 10% of NISP or even less), larger species had a greater proportion of thermally-altered specimens than the small ones.

Fig. 5.3 (a) Cutmarks and a sawed furrow running along the bone perimeter in a distal metapodial of *B. dichotomus*. (b) *O. bezoarticus* vertebra: cutmarks in the spinous process

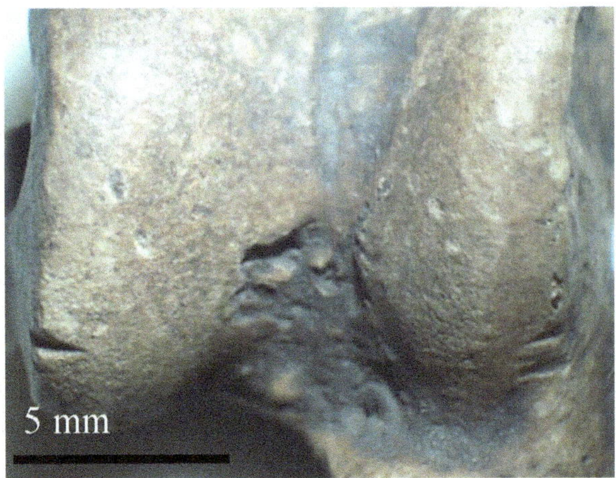

Fig. 5.4 Astragalus of *H. hydrochaeris* with cutmarks

For guanaco, the amount of burnt bones from the limbs such as tarsals, carpals or metapodials was quite remarkable. Alternatively, in the case of *O. bezoarticus*, a considerable amount of phalanges and astragali were also altered by fire.

5.3.2.2 Small Fauna

In addition to large mammals and birds, the inhabitants of Cañada Honda also exploited small fauna. Evidence of bird consumption of several species were the cutmarks located on the articular end of an *A. georgica* humerus, the distal end of a *P. olivaceus* femur, a femur of *R. rufescens*, and a coracoid of *D. bicolor*. A tibio-tarsus identified as Tinamidae showed a deep cutmark, as well as a fracture resulting from anthropic activity.

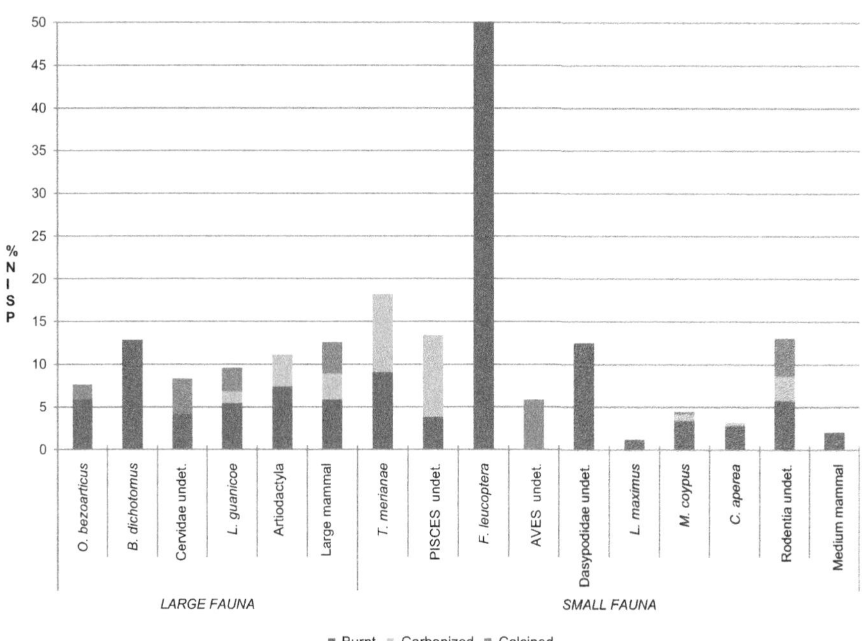

Fig. 5.5 Thermally-altered specimens expressed in %NISP (large and small fauna)

Rodentia was the order that shows more evidence of meat exploitation, particularly represented in cutmarks and scraping marks, both alterations thoroughly considered in previous papers (Escosteguy and Salemme 2012; Escosteguy et al. 2012). *M. coypus* yielded the largest diversity and quantity of marks, on either mandible, humerus (Fig. 5.6), radius, ulna, femur, tibiae, or third and fourth metatarsal. Furthermore, two proximal ends of *L. maximus* femora showed cutmarks around the head and the major trocanter.

Several variables observed in the assemblage from Cañada Honda indicated that fractures were produced on green bone, particularly in the case of *C. aperea*, *L. maximus* and *M. coypus*.

Long bones of cavids showed helical fractures, mainly on femora and humeri. Considering the largest species of rodents—*L. maximus* and *M. coypus*—long bones had evidence of helical and single flat fractures, with sawed furrow running along the bone perimeter; some of these fractures had also associated cutmarks. Coypu (*M. coypus*) hemi-mandibles were intentionally fractured as well. *L. gymnocercus* was the only carnivore with evidence of human manipulation: a radius presented cutmarks typical of transversal fracture on green bone. Neither cutmarks nor green fractures were identified on fish.

Regarding heat evidence on small sized species (Fig. 5.5), *F. leucoptera* yielded the highest percentage of thermally altered bones (%NISP = 50), but the result may be significantly affected by sample size (NISP = 2). The remaining species were,

Fig. 5.6 Cutmarks on a humerus shaft of *M. coypus*

however, comparable to the large fauna. Fish, armadillos and rodents had similar global percentages although burning categories depicted different rates. Fish resulted in a larger number of carbonized bones whereas only a few specimens were burnt. In general, a considerable number of cranial bones were affected by fire. The single altered armadillo bone rated as burnt in the scale color, whereas undeterminable rodent bones showed similar proportions in the three thermal alteration categories. Reptile bones of *T. merianae* had a higher percentage of burnt and carbonized bones than large mammals. For rodents like *L. maximus*, *C. aperea* and *M. coypus*, the percentage of thermally-altered bones was low. The skeletal distribution of burnt bones had been thoroughly discussed on previous papers (Escosteguy 2011; Escosteguy and Salemme 2012; Escosteguy et al. 2012); nevertheless, it may be noted that in the case of *M. coypus*, the appendicular skeleton, particularly the hindlimb, showed the highest frequency of fire modifications, whereas for the axial skeleton, only two hemi-mandibles were affected. Concerning *C. aperea*, a few humeri, femora and tibiae were burnt. Finally, only the shaft of one *L. maximus* femur was also burnt (Escosteguy and Salemme 2012).

5.3.3 Taphonomic Modifications of Natural Origin

Different natural agents and processes affected bone remains (Fig. 5.7). Weathering (*sensu* Beherensmeyer 1978) is visible on large vertebrate bones, though in low degrees (1–2) or at most degree 3. *L. guanicoe* weathered bones showed the highest percentage (70%) of weathering, which indicates certain time of exposure.

The presence of roots etching was recorded in similar rates for all large vertebrates (between 40 and 50%), meaning a quick burial and an active soil. Less important were the effects of rodent and carnivore action; the former left their teeth marks particularly on *H. hydrochaeris* bones (18%) and carnivores

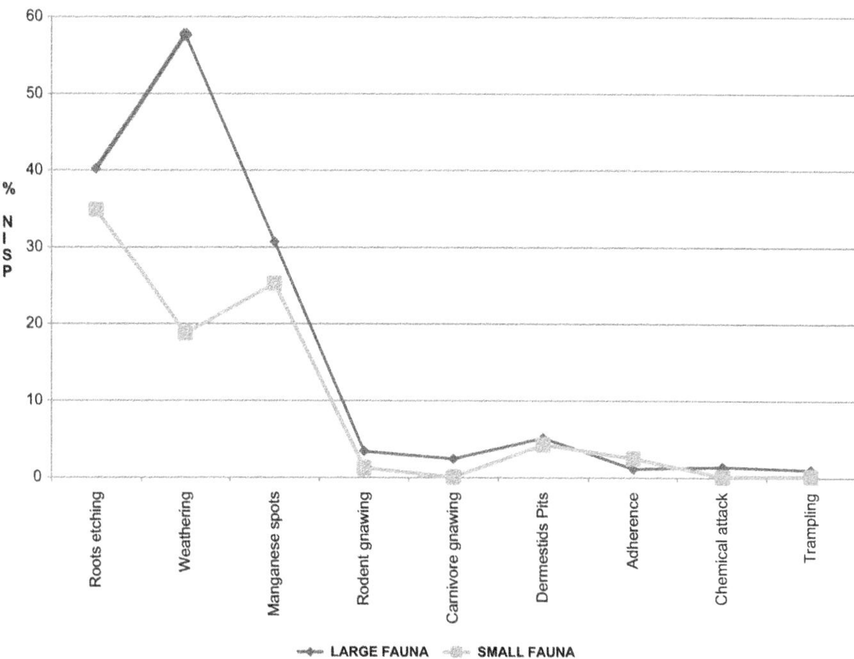

Fig. 5.7 Modifications of natural origin on bones of small and large fauna, in average

modified *O. bezoarticus* and *L. guanicoe* bones more frequently (between 6.85 and 8.74%). Some pits due to dermestids as well as trampling marks were visible in low proportions in all taxa, with the exception of *H. hydrochaeris*, whose bones were altered by trampling in the 18% of the NISP.

Regarding damage produced by inorganic substances, manganese spots were recorded in high proportion, being *R. americana* and the family Cervidae the taxa most frequently affected (between 44 and 50%). Chemical attacks and some sediment adherences were also present in a few bones, reaching very low percentages: the former comprised 1.41% of the NISP and the latter, just 1.14%.

A similar tendency in the case of natural modifications was observed in the small fauna assemblage, except for weathering, which was less important than root etching and manganese spots. Bones from different species of birds were affected by root etching in higher numbers (between 30 and 100% of NISP). On the other hand, species whose bones were widely modified by roots were *C. aperea*, *G. cuja*, and *F. geoffroyi*. The last two were also extensively weathered and altered by manganese oxide.

The rest of taphonomic modifications presented much lower percentages, being dermestid pits the most numerous (4.11%). Bird species were the most intensively modified taxa.

A comparison between small and large fauna regarding natural alterations is shown in Fig. 5.7. It is remarkable that both categories seem to have been affected by taphonomic agents and processes in a similar way, yielding roots etching, weathering and manganese spots as the most representative alterations. The remaining agents had a rather lower impact on larger and smaller vertebrate bones. The main difference between the two groups is the rate of weathering; this suggests a faster burying of small bones or a better protection by vegetation.

As it was mentioned above, natural modifications sometimes have shaded cultural marks on bone surfaces. The action of different taphonomic agents is coherent with the wet environment that surrounds Cañada Honda; the roots of the abundant vegetation could have altered bones whereas the action of water could have created the conditions for manganese oxide. On the other side, the action of carnivores on large and small fauna, could be explained as the capability of some species (as Geoffroy's cat) to destroy prey elements of small fauna (Álvarez et al. 2011). An agent like this could have generated, then, an important bias on the faunal assemblage.

5.4 Discussion

The geomorphologic configuration of the studied environment began *ca.* 3500 years ago, when the río Paraná delta emerged and the area turned into a fluvial environment where aquatic fauna like coypu, marsh deer and freshwater fishes are associated (Loponte 2007). Therefore, the area became favorable for human settlement (Lanzelotti et al. 2011).

In Cañada Honda Locality, technological developments show that people were well equipped to interact in different ways with the environment by the Late Holocene. They had bone tools, pottery vessels of different sizes and lithic artifacts that could have offered new possibilities to obtain more benefits from the exploitation of faunal and plant resources. As it was proposed by other authors, the economic strategies of diversification and intensification were related to other cultural developments as the intra- and inter-regional exchange, a demographic increase and longer or recurrent occupation of the same places (Quintana et al. 2002; González 2005; Quintana and Mazzanti 2014; Stoessel and Martínez 2014; among others). Some of these issues could be considered for this locality, as the presence of lithic raw materials from distant places and ritual practices—as human burials show—which highlight the social importance of this landscape (Migale and Bonaparte 2008).

The archaeofaunal assemblage here analyzed suggests different uses of vertebrates. They would have benefited from the wide diversity of species, both large and small, likely available all the year round and coming from terrestrial and aquatic environment. The zooarchaeological study of the Cañada Honda Locality revealed the exploitation of riverine fauna like ducks, whistling ducks, coots and cormorants, freshwater fishes, coypu, capybara and marsh deer. Other species from inland environments were intensively exploited as well: guanaco, ñandú, plains viscacha, pampas fox, red-winged tinamou and other specimens of Tinamidae. All the taxa recorded were available in the lowlands during the Late Holocene (Acosta

2005; Loponte 2007; Arrizurieta et al. 2010; Musali et al. 2013; Acosta and Mucciolo 2014). Likewise, some species as guanaco and Pampean deer could have been hunted in higher regions of the Undulate Pampa rather than in the lowlands (Politis et al. 2011; Acosta and Mucciolo 2014).

The analysis presented here was mainly devoted to the evidence of prey manipulation, supplementing the study of the skeletal parts of each taxon published elsewhere (Salemme 1987; Escosteguy 2011, 2012; Escosteguy and Salemme 2012; Escosteguy et al. 2012). The presence of certain marks and fractures on specific skeletal parts and bone sections is assumed as the result of faunal processing (butchering, skinning, filleting, etc.). Nevertheless, it is quite remarkable that direct evidence of cultural activity is only visible in very low rates. Partially, this rate might be influenced by the action of some natural agents (roots, manganese, etc.) which could have shaded anthropogenic evidences.

Cutmarks and fractures on large vertebrates are evidence of an integral use of carcass. Cutmarks were absent on the bones of the largest bird, *R. americana*, but fractures could be interpreted as the result of marrow consumption, as it was registered in other archaeological sites of the Pampean region (Acosta 2005; Salemme and Frontini 2011; Álvarez 2015). Likewise, the consumption of rheids' meat, smooth parts, feathers, skin, and tendons from legs was documented by travelers of the nineteenth century (Salemme and Frontini 2011). This species do not inhabit wetlands; therefore, ñandú was probably hunted on the steppes adjacent to the Undulate Pampa (Acosta 2005), since it was registered in other sites of Northeastern Buenos Aires province, though in limited rates.

There is evidence of integral exploitation on the family Artiodactyla; both, the appendicular and the axial skeleton presented different kind of marks (Escosteguy 2012). All the species within this family offered a large amount of meat, marrow and grease compared to small vertebrates. In addition, bones were employed as raw material to perform tools.

Instead, the stout capybara rodent had just one tarsus bone cut-marked, may be as the result of dismembering related to the skin acquisition (Escosteguy 2011).

Anthropic fractures and cutmarks were also recorded on small vertebrates, although some differences were observed. Most large prey and rodents like *L. maximus* and *M. coypus* (Escosteguy 2011; Escosteguy and Salemme 2012; Escosteguy et al. 2012) have been dismembered, skinned and butchered, whereas the small preys were processed in an alternative fashion and are devoid of marks. Both rodents—particularly coypu—have a smooth fur, which has been appreciated by Pampean people since pre-Hispanic times (Escosteguy 2011, 2014). Furthermore, the meat was consumed and the bones could have been crafted as tools (Escosteguy et al. 2012).

Another mammal with evidence of anthropic marks was the Pampean fox, which is one of the species with a thick fur in the area. Val and Mallye (2011) documented cutmarks on the experimental manipulation of carcasses of similar carnivores, particularly on radius, compatible with the example documented in the Cañada Honda assemblage. Consequently, it could be postulated that *L. gymnocercus* was hunted for its fur.

In the case of birds, the presence of cutmarks on femur, tibio-tarsus and coracoids would be pointing to dismembering flesh-rich sections for consumption, but also bones could be transformed as instruments and feathers used for personal ornaments. People who inhabited Cañada Honda exploited birds from terrestrial and aquatic habits; hence it could be expected different hunting strategies to acquire these preys. A similar archaeological record in the río Salado Depression includes, as well, avian species from both environments with anthropogenic evidences (González 2005). However, very low rates of bird bones with scarce anthropic actions on them have been informed from other sites of Northeastern Buenos Aires province. In fact, Acosta (2005)—considering ethnographic data—postulated the taboo on bird consumption.

Regarding thermal alterations, most of the large species were affected by fire in different degrees. On the opposite, small species presented less thermal damage, probably due to a different cooking strategy. Medium or small rodents were probably hardly dismembered to be later cooked in pottery containers. The high frequency of ceramics found in the site (Lanzelotti and Acuña 2010) would reinforce this idea.

Large prey could have been dismembered for transportation, then cooked on open fires or hearths or, at least, their bones tossed to the fire. Alternatively, Acosta and Mucciolo (2014) proposed that whole deer carcasses were transported to the camps to be intensively butchered; this practice would thus have generated the destruction of the axial skeleton. Cooking in a vessel would have maximized meat consumption. It could have been the case for deer at Cañada Honda. Likewise, for guanaco, particularly when it is abundantly represented by hind limbs and autopodials—as it happened in Cañada Honda—it was suggested its introduction to the site partially attached to the skin (Salemme 1987; Acosta and Mucciolo 2014).

Some taxa, like the small rodent *C. aperea* and fish in general, did not yield any cutmarks or fractures related to cultural activity, and very few remains presented thermal alterations. Nevertheless, their exploitation by hunter-gatherers is proposed on a different basis. To explain the significant accumulation of guinea pig bones, catastrophic events leading to their aggregation or the intervention of other natural agents such as carnivores or raptorial birds were discarded by the analysis of skeletal parts and marks on bone surfaces. The ethological habits of *C. aperea* might have contributed to the contamination of the record by the natural deposition of bones (Acosta et al. 2004); their accumulations, however, tend to be much smaller. In addition, data from archaeological sites in areas such as the "Bajíos Ribereños" may support the idea of a cultural origin (Acosta 2005; Acosta and Pafundi 2005; among others). Since thermal alteration was scarce and cutmarks were not visible, it could also be postulated that their meat was cooked in pottery vessels (Escosteguy 2011; Escosteguy and Salemme 2012); boiled carcasses would further explain the absence of cutmarks (Acosta and Pafundi 2005).

The presence of cutmarks on fish is unusual in the archaeological sites of the area. However, some assemblages showed green fractures on pectoral rays. Despite being considered the result of anthropic activity, the recovery context should be evaluated to rule out the action of carnivores or trampling. The fish bones from Cañada Honda did not present any cultural mark; however the presence of burnt bones (especially crania) could be thought as the consequence of discarding fish heads on burning fires (Acosta and Musali 2002).

Taphonomic damages might sometimes make difficult the recognition of cutmarks. Agents such as carnivores and rodents could have altered the garbage left by Late Holocene people, though they could not be considered the main collector of bones, since the frequency of tooth marks in the assemblage was quite low. Moreover, the different taphonomic agents and process of natural origin should have altered both types of animals in the assemblage (small and large vertebrates) in similar ways.

Our results show some coincidences with the work of other researchers. Acosta and Mucciolo (2014) described the presence of *O. bezoarticus* in different archaeological sites in the north-eastern portion of the Humid Pampas (Buenos Aires and Entre Ríos provinces). They observed that this species occupied the wetlands of the lower río Paraná, including flooded areas like the 'Bajíos ribereños', as well as higher and open lands with better drainage systems like the Undulate Pampa (Acosta and Mucciolo 2014). Therefore, the capture of this prey would have been possible in both inland and fluvial environments. The chorology of this species could have overlapped with the distribution of marsh deer (Politis et al. 2011), which would explain their co-occurrence in the assemblages.

Against expectation, capybara—a resource which may have been a potential prey due to its energetic capacity and low capture costs- was absent or very rare in numerous sites in the Pampean region, in particular in the study area. As regards to the human role in the accumulation and modification of the capybara, it was proposed that this large rodent was only occasionally exploited, perhaps being considered 'taboo' for Pampean people during the Late Holocene (Salemme 1987; Acosta 2005). The Cañada Honda zooarchaeological record would confirm this as it appeared that capybara was present in the area but people would not have hunted it.

The analysis presented here indicated the integral treatment of both small and large species. As well as the edible and palatable products from large prey (weighing between 40 and 150 kg), hunter gatherers made use of antlers and bones as raw material to manufacture instruments (Salemme 1987; Pérez Jimeno 2004; Escosteguy 2012). For small and medium vertebrates, a number of applications may be proposed; rodents were used for food, leather for making shelters and clothing, and bones as a source of raw material; for birds, meat, bones, feathers and eggs; for fish, flesh, oil, fish bones for tools or even fishmeal, as mentioned in historical sources (Acosta 2005; Musali et al. 2013).

In summary, the results from Cañada Honda Locality add new information to the intensification and diversification models (in the sense of Stoessel and Martínez 2014) for the diets of the hunter-gatherer groups of the Late Holocene in the Pampas, particularly in the northeastern Humid Pampas. External factors like environmental changes or demographic pressure would have encouraged people to optimize benefits from the available species, its direct exploitation or through exchange.

Acknowledgements The authors are indebted to the staff of the Museo de Ciencias Naturales "Carlos Ameghino" where the studied collection is deposited; Yolanda Davies (MACN Bernardino Rivadavia-CONICET) helped us with bird bone identification. Dr. David Flores and Dr. Darío Lijtmaer allowed us to consult faunal collections from División de Ornitología and Mastozoología (MACN Bernardino Rivadavia-CONICET). The map was drawn by Ramiro López (CADIC-CONICET) and Dr. Sonia Lanzelotti supplied the landscape photographs. Dr. Aixa Vidal and Dr. Jorge Rabassa helped with the translation and the two reviewers encourage us, with their comments, to improve the manuscript. A preliminary version of this paper was presented at the 12th ICAZ (2014), during the Session "Neotropical Zooarchaeology", co-organized by the editors of this volume, to whom we are also indebted. This research was funded by PICT 2013-0411 and UBACyT 20020130100134BA. The authors are the solely responsible for mistakes and/or omissions in the ideas presented here.

References

Acosta A (2005) Zooarqueología de cazadores-recolectores del extremo nororiental de la provincia de Buenos Aires (Humedal del río Paraná inferior, Región Pampeana, Argentina). PhD thesis, Universidad Nacional de La Plata

Acosta A, Mucciolo L (2014) Paisajes arqueofaunísticos: distribución y explotación diferencial de ungulados en el sector centro-oriental de la región pampeana. Arqueología 20(2):243–261

Acosta A, Musali J (2002) Ictioarqueología del Sitio La Bellaca 2 (Pdo. de Tigre, Pcia. de Buenos Aires). Informe preliminar. Intersec Antropol 3:3–17

Acosta A, Pafundi L (2005) Zooarqueología y tafonomía de Cavia aperea en el humedal del Paraná inferior. Intersec Antropol 6:59–74

Acosta A, Loponte D, Durán S et al (2004) Albardones naturales vs. culturales: exploraciones tafonómicas sobre la depositación natural de huesos en albardones del nordeste de la provincia de Buenos Aires. In: Martínez G, Gutiérrez G, Curtoni R, Berón M, Madrid P (eds) Aproximaciones Contemporáneas a la Arqueología Pampeana, Perspectivas teóricas, metodológicas, analíticas y casos de estudio. Universidad Nacional del Centro de la Provincia de Buenos Aires, Olavarría, pp 77–91

Álvarez MC (2015) Utilización de Rhea americana (Aves, Rheidae) en el sitio Paso Otero 4 (partido de Necochea, región pampeana). Archaeofauna 24:53–65

Álvarez MC, Kaufmann C, Massigoge A et al (2011) Bone modification and destruction patterns of leporid carcasses by Geoffroy's cat (Leopardus geoffroyi): an experimental study. Quat Int 278:71–80

Andrews P (1990) Owls, caves, and fossils. University of Chicago Press, Chicago

Arrizurieta MP, Buc N, Mazza B et al (2010) Nuevos aportes a la arqueología del sector continental del Humedal de Paraná Inferior. In: Bárcena J, Chiavazza H (eds) XII Congreso Nacional de Arqueología Argentina. Mendoza, October 2010. Arqueología Argentina en el Bicentenario de la Revolución de Mayo, Tomo V. Facultad de Filosofía y Letras, Universidad Nacional de Cuyo, Cuyo, pp 1793–1797

Behrensmeyer AK (1978) Taphonomic and ecologic information from bone weathering. Paleobiology 4(2):150–162

Binford LR (1981) Bones: ancient men and modern myths. Academic Press, New York

Bonaparte J (1951) Nota preliminar de un paradero aborigen en Cañada Honda (Baradero). In: Apuntes de difusión científico-cultural. Museo Popular de Ciencias Naturales "Carlos Ameghino"

Bonfils C (1962) Los suelos del Delta del Río del Paraná. Factores generadores, clasificación y uso. Rev de Invest Agraria, INTA 6:3

Escosteguy PD (2011) Etnoarqueología de nutrieros. Una propuesta metodológica aplicada al registro arqueológico de la Depresión del Salado y del Noreste de la provincia de Buenos Aires. PhD thesis, Universidad de Buenos Aires

Escosteguy PD (2012) Los cérvidos de la localidad arqueológica Cañada Honda (Baradero, Buenos Aires). Análisis preliminar. Comechingonia 16(2):163–168

Escosteguy PD (2014) Estudios etnoarqueológicos con cazadores de coipo de Argentina. Rev Antípoda 20:145–165

Escosteguy PD, Salemme MC (2012) Butchery evidence on rodent bones from archaeological sites in the Pampean Region (Argentina). In: Lefèvre C (ed) Proceedings of the general session of the 11th ICAZ international conference (Paris, 23–28 August 2010), BAR international series 2354. Archaeopress, Oxford, pp 227–237

Escosteguy PD, Salemme MC, González MI (2012) Myocastor coypus ("coipo", Rodentia, Mammalia) como recurso en los humedales de la Pampa bonaerense: patrones de explotación. Rev del Museo de Antropol 5:13–30

Fisher J (1995) Bone surface modifications in zooarchaeology. J Archaeol Method Theory 2 (1):7–68

Gómez G (2000) Análisis tafonómico y paleoecológico de los micro y mesomamíferos del sitio arqueológico de Arroyo Seco 2 (Buenos Aires, Argentina) y su comparación con la fauna actual. PhD thesis, Universidad Complutense de Madrid

González MI (2005) Arqueología de alfareros, cazadores y pescadores pampeanos. Sociedad Argentina de Antropología, Buenos Aires

Grayson D (1984) Quantitative zooarchaeology. Academic Press, Orlando

Gutiérrez M (2004) Análisis tafonómicos en el Área Interserrana (provincia de Buenos Aires). PhD thesis, Universidad Nacional de La Plata

Lanzelotti S, Acuña G (2010) A 60 años del descubrimiento de Cañada Honda: interpretaciones y reinterpretaciones de su cerámica. In: Berón M, Luna L, Bonomo M, Montalvo C, Aranda C, Carrera Aizpitarte M (eds) Mamül Mapu: pasado y presente desde la arqueología pampeana, Tomo II. Editorial Libros del Espinillo, Ayacucho, pp 293–307

Lanzelotti S, Bonaparte J (2009) Contexto geoestratigráfico y procesos de formación del registro arqueológico en Cañada Honda: apuntes para su discusión y abordaje. In: Fucks E, Deschamps C, Silva CG, Schnack E (eds) IV Congreso Argentino de Cuaternario y Geomorfología, XII Congresso da Associação Brasileira de Estudos do Quaternário, II Reunión sobre el Cuaternario de América del Sur, September 2009. Universidad Nacional de La Plata, La Plata, pp 247–258

Lanzelotti S, Politis G, Carbonari J et al (2011) Aportes a la cronología del Sitio 1 de Cañada Honda (partido de Baradero, provincia de Buenos Aires). Intersec Antropol 12:355–361

Loponte D (2007) La economía prehistórica del norte bonaerense (Arqueología del humedal del Paraná inferior, Bajíos Ribereños meridionales). PhD thesis, Universidad Nacional de La Plata

Lyman RL (1994) Vertebrate taphonomy. Cambridge University Press, Cambridge

Lyman RL (2008) Quantitative paleozoology. Cambridge University Press, Cambridge

Martínez G, Gutiérrez M (2004) Tendencias en la explotación humana de la fauna durante el Pleistoceno final y Holoceno en la Región Pampeana (Argentina). In: Mengoni Goñalons G (ed) Zooarchaeology of South America, BAR international series, vol 1298, pp 81–98

Mengoni Goñalons G (1999) Cazadores de guanacos de la estepa patagónica. Sociedad Argentina de Antropología, Buenos Aires

Migale L, Bonaparte J (2008) Arqueología de Cañada Honda y Río Areco. Baradero, Buenos Aires. Fondo Editorial Mercedes, Mercedes

Miotti L, Salemme MC (1988) De fracturas óseas: arqueológicas y modernas. Rev Estudios Regionales CIDER 2:17–26

Musali J, Feuillet Terzaghi MR, Sartori J (2013) Análisis comparativo de conjuntos ictioarqueológicos generados por cazadores-recolectores durante el Holoceno tardío en la baja Cuenca del Plata (Argentina). Cuadernos del Instituto Nacional de Antropología y Pensamiento Latinoamericano—Series Especiales 1(1):211–225

Pérez Jimeno L (2004) Análisis comparativo de dos conjuntos de artefactos óseos procedentes de la llanura aluvial del Paraná y la pampa bonaerense. In: Martínez G, Gutiérrez M, Curtoni R, Berón M, Madrid P (eds) Aproximaciones Contemporáneas a la Arqueología Pampeana, Perspectivas teóricas, metodológicas, analíticas y casos de estudio. Universidad Nacional del Centro de la Provincia de Buenos Aires, Olavarría, pp 319–333

Pérez Ripoll M (2005/2006) Caracterización de las fracturas antrópicas y sus tipologías en huesos de conejo procedentes de los niveles gravetienses de la Cova de les Cendres (Alicante). MUNIBE (Antropologia-Arkeologia) 57(1):239–254. Sociedad de Ciencias Aranzadi Zientzi Elkartea

Politis G, Prates L, Merino M et al (2011) Distribution parameters of guanaco (Lama guanicoe), pampas deer (Ozotoceros bezoarticus) and marsh deer (Blastocerus dichotomus) in Central Argentina: archaeological and paleoenvironmental implications. J Archaeol Sci 38:1405–1416

Quintana C (2007) Marcas de dientes de roedores en huesos de sitios arqueológicos de las Sierras de Tandilia, Argentina. Archaeofauna 16:185–191

Quintana C, Mazzanti D (2014) La emergencia de la diversificación de la caza en las Sierras de Tandilia Oriental durante el Holoceno tardío final. Comechingonia 18(2):41–64

Quintana C, Valverde F, Mazzanti D (2002) Roedores y lagartos como emergentes de la Diversificación de la Subsistencia durante el Holoceno Tardío en Sierras de la Región Pampeana Argentina. Lat Am Antiq 13(4):455–473

Salemme MC (1987) Paleoetnozoología del sector bonaerense de la región Pampeana, con especial atención a los mamíferos. PhD thesis, Universidad Nacional de La Plata

Salemme MC (1990) Zooarchaeological studies in the Humid Pampas, Argentina. Quat S Am Antarct Peninsula 6(1988):309–335

Salemme MC, Frontini R (2011) The exploitation of Rheidae in Pampa and Patagonia (Argentina) as recorded by chroniclers, naturalists and voyagers. J Anthropol Archaeol 30:473–483

Salemme MC, Escosteguy PD, Frontini R (2012) La fauna de porte menor en sitios arqueológicos de la región pampeana, Argentina. Agente disturbador vs. recurso económico. Archaeofauna 21:163–185

Stoessel L, Martínez G (2014) El proceso de intensificación en la transición pampeano-patagónica oriental. Discusión y perspectivas comparativas con regiones aledañas. Comechingonia 18 (2):65–94

Val A, Mallye JB (2011) Small carnivore skinning by professionals: skeletal modifications and implications for the European Upper Palaeolithic. J Taphonomy 9:221–243

Space Use Patterns and Resource Exploitation of Shell Middens from the Río de La Plata Coast (*ca.* 6000–2000 Years BP), Uruguay

6

Laura Beovide, Sergio Martínez, and Walter Norbis

6.1 Introduction

The oldest records of human presence in the *Río de la Plata* estuary are Late Pleistocene (Suárez and López 2003), but is *circa* the middle Holocene when the archaeological signal became more abundant. They were hunter-fisher-gatherer societies that have incorporated into their economy the raising of different cultigens and pottery production (Beovide 2010). This pottery-making societies practiced small-scale horticulture incorporating *Zea mays* (*ca.* 3000 ^{14}C years BP) within their set of cultigens (Beovide 2011a).

The study of pre-Hispanic archaeological sites associated with shell middens in the Uruguayan coast of the *Río de la Plata* was developed recently, just in the last decade (Beovide and Martínez 2014). Two main reasons concur to explain this late development: lack of systematic research in Archaeozoology in this geographic area, and the difficulty of distinguishing between the anthropic and natural shelly deposits because both types share the same species of mollusks (Beovide and Martínez 2014).

We identified eight shell deposits of anthropic origin in the lower part of the *Santa Lucia* river basin, at the left margin of the *Río de la Plata* (Fig. 6.1). Nevertheless, using multiple lines of evidence (Actualistic, Taphonomic,

L. Beovide (✉)
Centro de Investigación Regional Arqueológica y Territorial-DICYT-MEC, Montevideo, Uruguay
e-mail: lbeovide@dicyt.gub.uy

S. Martínez
Departamento de Evolución de Cuencas, Facultad de Ciencias, UdelaR, Montevideo, Uruguay
e-mail: smart@fcien.edu.uy

W. Norbis
Facultad de Ciencias, Instituto de Biología, Departamento de Biología Animal, UdelaR, Montevideo, Uruguay
e-mail: wnorbis@fcien.edu.uy

M. Mondini et al. (eds.), *Zooarchaeology in the Neotropics*,
DOI 10.1007/978-3-319-57328-1_6

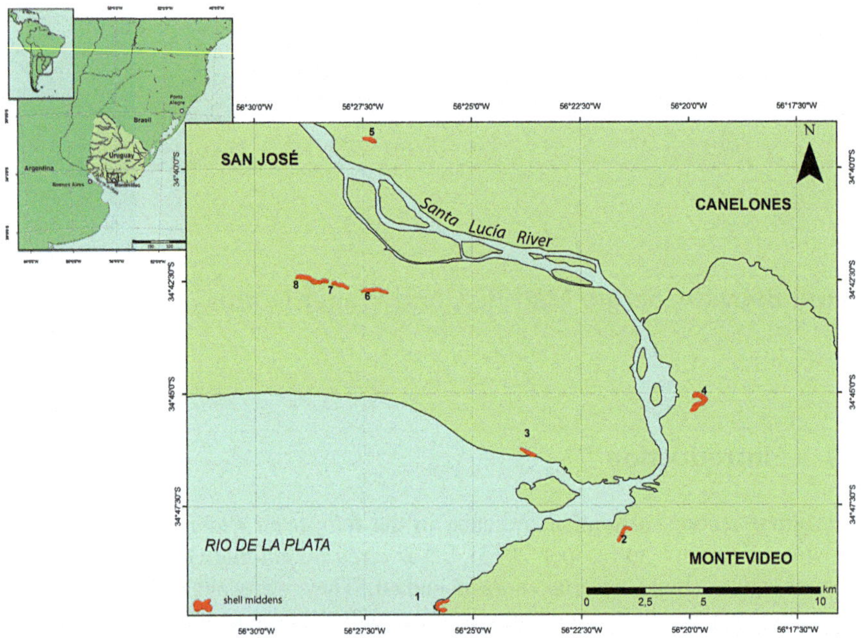

Fig. 6.1 Location of the archaeological shell middens (Santa Lucia basin). References: 1. Punta Espinillo, 2. Cañada de las Conchas, 3. Km 26, 4. Dianova, 5. Gambé, 6. Colonización (6), 7. Colonización (7), 8. Colonización (8)

Ethnoarchaeological, and Geoarcheological studies) we managed to discriminate one type from another (Beovide 2011b, 2014; Beovide et al. 2014a, b).

They are situated along old coastlines developed since the climatic optimum, and are dated from 6000 to 2000 ^{14}C years BP.

The first shell middens of the Santa Lucia river basin were made up during the local maximum sea level, i.e. around 7000 a 6000 years BP (Cavallotto et al. 2004; Violante and Parker 2004). To some authors, the sea level was in this moment around +5 to +10 m amsl (Urien 1970), but to others it reached up to +5 to +6.5 m amsl (Cavallotto et al. 2004). There is also not agreement regarding the existence of a posterior oscillating (Bracco et al. 2005, 2010, among others) or gradual falling of the sea level (Martínez and Rojas 2013, among others). The Holocene paleo-environments records of the study area are described in several papers (e.g. Bracco et al. 2005, 2010; Iriarte 2006; Martínez and Rojas 2013). The presence of shell middens in the lower *Santa Lucía* basin is concordant with an scenario of a changing environment from relatively hot and humid times to more arid ones (Beovide 2009, 2011a).

Between *ca.* 6000 ^{14}C years BP and 2000 ^{14}C years BP there was a development of an estuary, and lagoons formed as a consequence of the sea level falling (Beovide 2011b). Figure 6.2 shows the location of the shell middens in such a landscape.

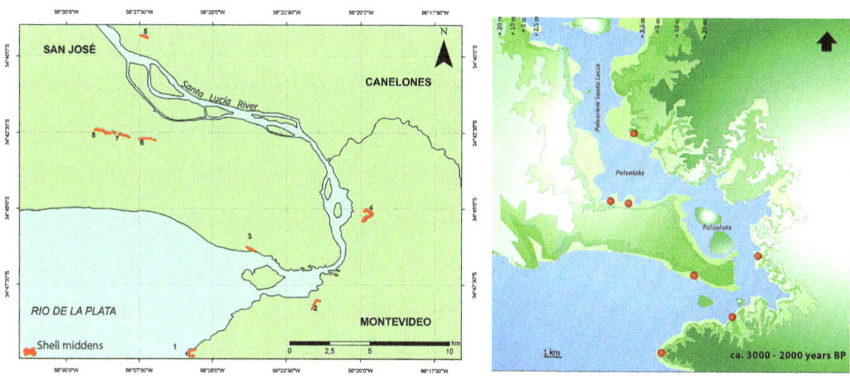

Fig. 6.2 *Santa Lucia* river and shell middens. Current (*left*) and *Santa Lucia* paleo-river model (*right*, based Beovide 2011b)

6.2   Methodology

First, we did random test regarding the spatial distribution of the shell middens (sites). A nearest neighbour index analysis (Clark and Evans 1954; Krebs 1999) was done, in order to know if the shell middens are distributed by chance, in a uniform or disperse way or in an aggregate one. Then, we analyzed the probable causes of the distribution through the study of the available resources (Hodder and Orton 1990).

The shell middens are at old paleocoasts or over rocky points, between +4.5 and +10 m amsl. About ten test pits of 1–1.5 m in each midden were done. In two of them (*Gambé* and *Colonización*) five excavations were done too. In average *ca.* 25 m^2 and *ca.* 60 m^2 were excavated in a systematic and controlled way in *Gambé* and *Colonización* 6 respectively, following the stratigraphic units and using SIG to register spatially materials and structures. We collected two bulk samples with a weight of 2.5 kg and a volume of around 30 × 30 × 30 cm each. Obtained radiocarbon dates and related data are shown in Table 6.1.

The molluscs present in the shell middens and the other resources (plants, terrestrial and aquatic animals, minerals) were identified to the lowest taxonomic possible level.

We constructed two data matrices, one of them regarding the composition of each sample in grams: archaeological material, mollusks concentration, sediment, and larger clasts; and other matrix taking into account the Minimum Number of Individuals (MNI) and the Number of Remains (NR), according to Giovas (2009) and Mason et al. (1998). To obtain the MNI we counted separately right and left valves of a given species, and the most abundant of them, plus the articulated specimens, was considered the final number. The NR surge by counting every taxonomically identified unit in the sample.

The taphonomical analysis of the sites was published elsewhere (Beovide 2011b, 2014; Beovide et al. 2014a, b). In these works disarticulation: disarticulation, shell

Table 6.1 Radiocarbon and calibrated dates for shell midden and additional data

Site	Exc.	Level	^{14}C (years BP)	Calibrated dates: (68%) 1σ	Material/species	Lab./Nr.	References
Cañada de las conchas	Test pits 6	3	6651 ± 33 (AMS)	7508–7566 cal BP, 5587 ± 29 cal BC	*Mactra isabelliana*	AA/104633	Beovide and Campos (2014)
Dianova	Test pits 3	6	4977 ± 32 (AMS)	5667–5739 cal BP, 3753 ± 36 cal BC	*Ostrea equestris*	AA/104634	Beovide and Campos (2014)
Punta Espinillo	Test pits 1	3	3088 ± 31 (AMS)	3273–3354 cal BP, 1364 ± 40 cal BC	*Tagelus plebeius*	AA/104632	Beovide and Campos (2014)
Punta Espinillo	Test pits	40 cm	3790 ± 140	3989–4380 cal BP, 2235 ± 195 cal BC	Valves	URU 09	Bracco (1994)
Colonización	III	7	2744 ± 50 (AMS)	2797–2907 cal BP, 902 ± 55 cal BC	*Anomalocardia brasiliana*	AA/86685	Beovide (2011b)
Colonización	S4	7	2716 ± 36 (AMS)	2785–2851 cal BP, 968 ± 33 cal BC	*Erodona mactroides*	AA/86686	Beovide (2011b)
Gambé	III	7	2710 ± 60	2779–2879 cal BP, 879 ± 50 cal BC	*Erodonamactroides*	URU/309	Beovide et al. (2001)
Colonización	V	8	2699 ± 29 (AMS)	2777–2838 cal BP, 858 ± 30 cal BC	*Erodona mactroides*	AA/104636	Beovide et al. (2014a)
Colonización	IV	10	2620 ± 50	2718–2783 cal BP, 801 ± 32 cal BC	*Erodona mactroides*	URU/0551	Beovide (2011b)
Colonización	IV	15	2600 ± 50	2598–2762 cal BP, 730 ± 82 cal BC	*Erodona mactroides*	URU/0552	Beovide (2011b)
Gambé	S8	7	2409 ± 32 (AMS)	2382–2598 cal BP, 540 ± 108 cal BC	*Erodona mactroides*	AA/86684	Beovide (2011b)
Gambé	III	5	2365 ± 32 (AMS)	2354–2446 cal BP, 450 ± 46 cal BC	*Erodona mactroides*	AA/86683	Beovide (2011b)
Colonización	I	5	2310 ± 60	2211–2416 cal BP, 364 ± 102 cal BC	*Erodona mactroides*	URU/0310	Beovide et al. (2001)

weathering due to biological and physical-chemical factors, with weathering stages from 1 to 6 from fresh shell to its destruction at touch, fracture type, and the presence of holes were recorded (Kotzian and Simões 2006; Favier and Borella 2007; Martínez et al. 2006; Zuschin et al. 2003).

The richness and diversity are size sample dependent measures. Larger samples tend to show greater taxonomic richness that smaller under similar conditions of integrity. The sample rarefaction procedure (Mao tau) (Colwell et al. 2004) was applied to a matrix of presence-absence data from Table 6.2, to estimate species richness as a function of number of samples. Also we applied individual rarefaction procedure (Krebs 1999) (from NMI data in Table 6.2) to estimate the number of classes or taxa in samples having unequal frequencies which can then be compared with each other at the same level. The small mollusk *Heleobia* sp. was excluded in this last analysis because the abundance masks the rest of the species.

We considered richness (the number of molluscan species) and potential resource distribution (Binford 1980, 2001; Dincauze and Driver 2003; Gould 1978, among others). The potential resource distribution is determined relating the resources found in the shell middens with their potential places of origin within the regional resource structure, taking into account the environmental dynamics. Site Territorial Analysis (STA), defined as the areas habitually used for daily subsistence from given locations (Higgs and Vita-Finzi 1972).

In this sense, we studied the probable area for obtaining the mollusks. We established three ranges of potential presence of resources: 0–3 km, 0–10 km and more than 10 km) that represent a potential immediate supply, local or not, Franco (2004). In this case we considered our own previous studies about the regional structure of the resources and archaeological sites of the lower *Santa Lucía* basin (Beovide 2001, 2009, 2011b; Martínez et al. 2006).

The distribution, predictability, density and seasonality of resources in relation to minerals, botanical, mollusks, fishes and mammals resources found in the shell middens (Table 6.4) was analyzed by means of factorial analysis of correspondences (Benzecri 1980; Greenacre 1984).

Correspondence analysis (CA) is a form of ordination that rearranges in this case a resources-by-variables matrix, to reflect its inherent order without prior grouping of the samples. It is reciprocal in the sense that scores for both variables and cases are jointly determined in the same ordination space by their weighted average. Scores of variables in CA ordination space mark their centers of cooccurrences, and relative positions of case scores reflect their variables compositions. Cases with similar CA scores have variables compositions, whereas variables with similar CA scores tend to co-occur. Thus, display of CA scores for variables and cases can be mutually compared to identify their joint associations. The procedure is efficient for presence-absence data because they will be well described by chi-square distances.

For all analysis, we used PAST version 2.17c, a free statistical software package (Hammer et al. 2001). In all tests, the significance level used was $p = 0.05$.

Table 6.2 Mollusk species found in the shell middens

Shell midden	Punta Espinillo				Cañada de las Conchas				Dtanova				Gantbé			
Samples	M4		M1xe		M1S6		M2xcc		MLD		M1xD		M42		M33x	
Weight	2500 g		2500 g		2500 g		2500 g		2500 g		2500 g		2500 g		2500 g	
NR/MNI	NR	NMI	NR	NMI	NR	NMI	NR	NMI	NR	NMI	NR	NMI	NR	NMI	NR	NMI
Erodona maaroides	0	0	0	0	0	0	0	0	0	0	60	25	2331	1186	713	386
Helobia sp.	420	420	264	264	618	618	150	150	0	0	0	0	###	79,900	2093	2093
Ostrea equestris	491	240	108	54	880	426	800	398	670	318	250	122	1	1	0	0
Tagelus plebeius	0	0	60	30	160	78	55	26	0	0	0	0	0	0	1	1
Anomahcar diabrasiliana	0	0	0	0	64	32	24	12	0	0	0	0	0	0	0	0
Maclra sp.	0	0	0	0	96	48	22	10	0	0	0	0	0	0	0	0
Gastropods	0	0	0	0	0	0	0	0	0	0	0	0	0	0	0	0
Mytilus edulis	470	226	468	234	65	30	0	0	36	18	0	0	0	0	0	0
Buccinanops deformis	6	6	6	6	0	0	0	0	0	0	0	0	0	0	0	0
Picatula gibbosa	0	0	12	6	14	7	0	0	0	0	0	0	0	0	0	0
Siphonaria lessoni	24	24	6	6	0	0	0	0	0	0	0	0	0	0	0	0
Mann whitney	p = 0.495; p > 0.05				p = 0.2551; p > 0.05				p = 0.9611; p = 0.05				p = 0.9666; p > 0.05			

Shell midden	Colonización 6		Colonización 7		Colonización 8		Km 26	
Samples	M25	M23	M1x7	M2x7	M1x8	M2x8	M126	M226
Weight	2500 g	2500 g	2500 g	2500 g	2500 g	2500 g	2500 g	2500 g
NR/MNI	NR	NMI	NR	NMI	NR	NMI	NR	NMI	NR	NMI	NR	NMI	NR	NMI	NR	NMI
Erodona maaroides	241	138	870	606	470	210	760	650	235	100	600	389	4400	2190	66	30
Helobia sp.	709	709	546	546	0	0	0	0	0	0	0	0	80	80	0	0
Ostrea equestris	1	1	0	0	0	0	0	0	0	0	0	0	0	0	0	0
Tagelus plebeius	1	1	1	1	0	0	0	0	0	0	0	0	0	0	0	0
Anomahcar diabrasiliana	1	1	1	1	0	0	0	0	0	0	0	0	0	0	0	0
Maclra sp.	1	1	0	0	0	0	0	0	1	0	0	0	0	0	0	0
Gastropods	1	1	1	1	0	0	0	0	0	0	0	0	0	0	0	0
Mytilus edulis	0	0	0	0	0	0	0	0	0	0	0	0	0	0	0	0
Buccinanops deformis	0	0	0	0	0	0	0	0	0	0	0	0	0	0	0	0
Picatula gibbosa	0	0	0	0	0	0	0	0	0	0	0	0	0	0	0	0
Siphonaria lessoni	0	0	0	0	0	0	0	0	0	0	0	0	0	0	0	0
Mann whitney	p = 0.5471; p > 0.05							

Table 6.3 Mollusks species grouped by site and relative distance between their living area and the sites analyzed

	Punta Espinilo			Cañada de las Conchas			Dianova			Gambé			Colonización 6			Colonización 7			Colonización 8			Km 26		
	>3	3–10	<10	>3	(3–10)	<10	>3	(3–10)	<10	>3	(3–10)	<10	>3	(3–10)	<10	>3	(3–10)	<10	>3	3–10	<10	>3	3–10	<10
Eradona mactroides	–	–	–	–	–	–	X	–	–	X	–	–	X	–	–	X	–	–	X	–	–	X	–	–
Heleobia sp.	X	–	–	X	–	–	–	–	–	X	–	–	X	–	–	–	–	–	–	–	–	X	–	–
Ostrea equestris	X	–	–	X	–	–	X	–	–	X	–	–	X	–	–	–	–	–	–	–	–	–	–	–
Tagelus plebeius	X	–	–	X	–	–	–	–	–	X	–	–	X	–	–	–	–	–	–	–	–	–	–	–
Anomalocardia brasiiana	–	–	–	X	–	–	–	–	–	–	–	–	X	–	–	–	–	–	–	–	–	–	–	–
Mactra sp.	–	–	–	X	–	–	–	–	–	–	–	–	X	–	–	–	–	–	–	–	–	–	–	–
Gastropod	–	–	–	–	–	–	–	–	–	–	–	–	X	–	–	–	–	–	–	–	–	–	–	–
Mytiks edulis	X	–	–	X			–	X	–	–	–	–	–	–	–	–	–	–	–	–	–	–	–	–
Buccinanops deformis	X	–	–	–	–	–	–	–	–	–	–	–	–	–	–	–	–	–	–	–	–	–	–	–
Plicatula gibbosa	X	–	–	X	–	–	–	–	–	–	–	–	–	–	–	–	–	–	–	–	–	–	–	–
Siphonaria lessoni	X	–	–	–	–	–	–	–	–	–	–	–	–	–	–	–	–	–	–	–	–	–	–	–

Reference: – (absence), X (presence)

Table 6.4 Predictability, potential distribution, density and seasonality and minerals, botanical, mollusks, fishes and mammals resources found in the shell middens

		Potential resource distribution			Potential resource predictability		Potential resource density		Seasonality (potential) resources	
		Source: 0–3 km	Source: 0–10 km	Source: greater than 10 km	Unpredictable	Predictable	Dense	Scarce	Annual	Seasonal
Mineral resources	Amphibolite	X	X	–	–	X	X	–	X	–
	Quartz	X	X	–	–	X	X	–	X	–
	Quartzite	X	X	–	–	X	X	–	X	–
	Granite	X	X	–	–	X	X	–	X	–
	Quartzite (Tajes)	–	X	–	–	X	X	–	X	–
	Silicified limestone	–	–	X	–	X	–	X	X	–
	Malachite	–	–	X	–	X		X	X	–
	Clay	X	–	–	–	X	X	–	X	–
	Sand	X	–	–	–	X	X	–	X	–
Botanical resources	*Ipomeabatata*	X	X	–	–	X	X	–		X
	Celtis tala	X	X	–	–	X	X	–	X	X
	Lupinus albescens	X	X	–	–	X	X	–	X	–
	Phaseolus vulgaris	X	X	–	–	X	X	–	–	X
	Syagrus romanzoffiana	X	X	–	–	X	X	–	–	X
	Cucurbita sp.	X	X	–	–	X	X	–	X	X
	Zea mays	X	X	–	–	X	X	–	X	X

(continued)

Table 6.4 (continued)

		Potential resource distribution			Potential resource predictability		Potential resource density		Seasonality (potential) resources	
		Source: 0–3 km	Source: 0–10 km	Source: greater than 10 km	Unpredictable	Predictable	Dense	Scarce	Annual	Seasonal
	Canna sp.	X	X	–	–	X	X	–	X	X
	Cyperaceae	X	X	–	–	X	X	–	X	–
Faunistic resources (mollusks)	*Erodona mactroides*	X	X	–	–	X	X	–	X	–
	Heleobia sp.	X	X	–	–	X	X	–	X	–
	Tagelus plebeius	X	X	–	–	X	X	–	X	–
	Anomalocardia brasiliana	X	X	–	–	X	–	X	X	–
	Ostrea equestris	X	X	–	–	X	–	X	X	–
	Mactra sp. *Mytilus edulis*	X	X	–	–	X	X	–	X	–
	Buccinanops deformis	X	–	–	X	–	–	X	–	X
	Plicatula gibbosa	X	–	–	X	–	–	X	–	X
	Siphonaria lessoni	X	–	–	–	X	–	X	X	–
	Gastropod	X	X	–	–	X		X	X	–

Faunistic resources (fishes)	*Pogonias cromis*	X	X	–	–	X	X		X	X
	Myliobatis sp.	X	X	–	–	X		X	X	–
	Shark (1)	X	X	X	X	–	–	–	–	X
	Shark (2)	–	X	X	X	–	–	–	–	X
	Shark (3)	–	X	X	X	–	–	–	–	X
Faunistic resources (mammals)	*Ozotoceros bezoarticus*	X	X	X	–	X	–	–	X	–
	Myocastor coypus	X	X	–	–	X	–	–	X	–
	Cavia aperea	X	X	–	–	X	–	–	X	–

Reference: – (absence), X (presence)

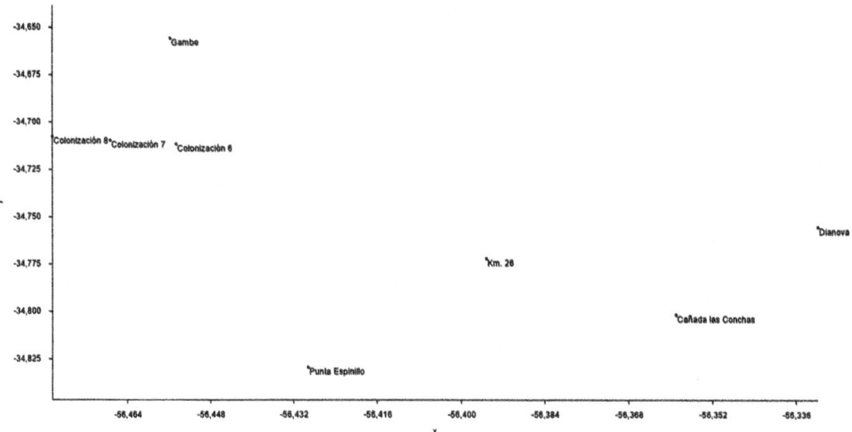

Fig. 6.3 Geographic coordinates and plot of the eight points where the shell middens associated to resource used were localized (see also Fig. 6.1)

6.3 Results

6.3.1 Shell Middens Spatial Distribution: Nearest-Neighbor Analysis

According to the nearest neighbour index analysis the distribution of the resources associated to the settlements, tends to be dispersed (non-random) ($R = 1.48$; $Z = 2.18$; $p = 0.029 < 0.05$). The location of the eight shell middens defined by their geographic coordinates are show in Fig. 6.3.

6.4 Mollusks Present in the Shell Middens

The richness (MNI and NR) and taxa present in each shell midden are detailed in Fig. 6.4 and Table 6.2. Differences between the samples taken within each shell midden are not significant, according to the Mann Whitney test (Sokal and Rohlf 1995), calculated for the samples with more than one species (Table 6.2).

The sample rarefaction accumulation curve (presence-absence data) showed a trend to increase to about 9–11 expected species (Fig. 6.5).

Two of the species accumulation curves (PE and CC) rises very rapidly in the beginning, and C6 accumulation curve showed a trend to increase with around five to six expected species, but with different number of individuals (Fig.6.6). DI accumulation curve was uniform with around two to three expected species. GA accumulation curve showed a lineal trend to increase with around two to three expected species and three species accumulation curves (C7, C8 and KM) not showed trends with around one expected species (Fig. 6.6). According to the

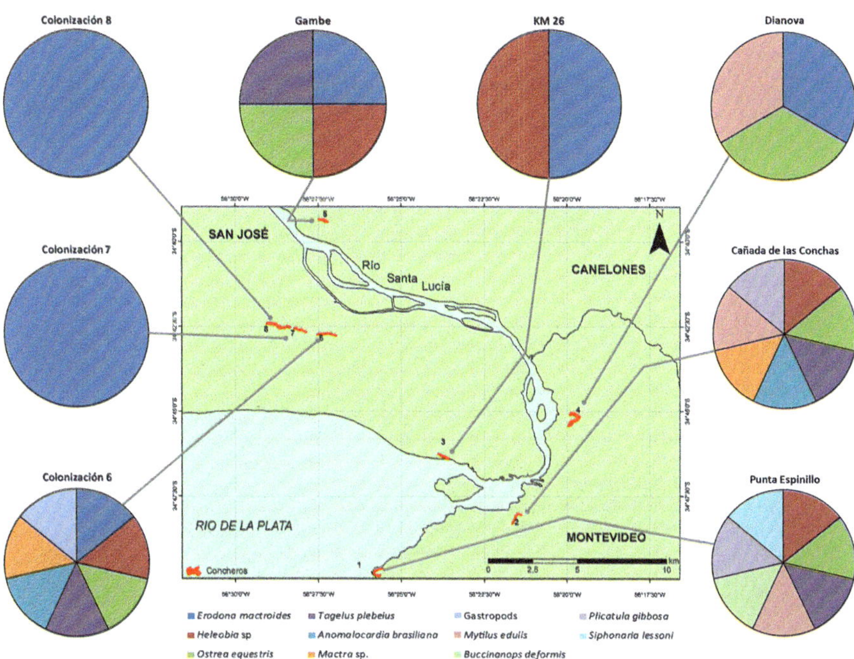

Fig. 6.4 Spatial distribution of the sites analyzed and proportion of mollusks species of the shell middens

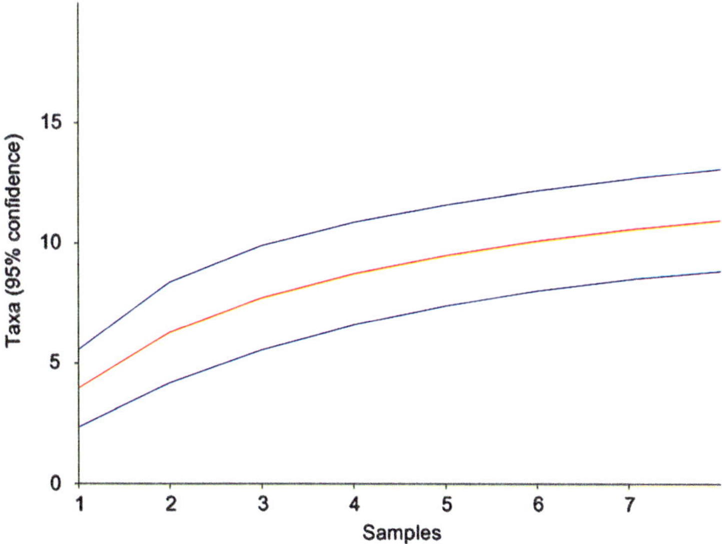

Fig. 6.5 Sample rarefaction accumulation curve (presence-absence data from Table 6.2). The *blue line* represents the 95% confidence interval for the curve, and samples in X-axis represents the sites analyzed

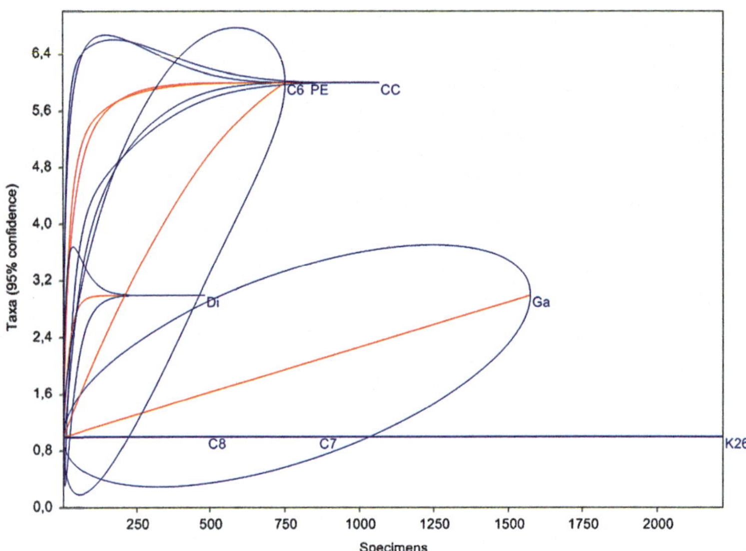

Fig. 6.6 Rarefaction curves without *Heleobia* sp. from the shell middens sites analyzed (see Fig. 6.1 for spatial distribution; PE = *Punta Espinillo*; CC = *Cañada de las Conchas*; KM = *Km 26*; DI = *Dianova*; GA = *Gambé*; C6 = *Colonización* (6); C7 = *Colonización* (7); C8 = *Colonización* (8)). The *blue line* represents the 95% confidence interval for the curves (Data from Table 6.2)

shape of the curves, we found a great variability in the number of individuals for the expected species richness in the shell middens sites analyzed.

All the mollusks suggests an estuarine to marginal marine environment.

The oldest shell midden (*ca.* 7000 a 6000 ^{14}C years BP, *Cañada de las Conchas*) has seven species (*Ostrea equestris, Tagelus plebeius, Anomalocardia brasiliana, Mactra* sp., *Mytilus edulis, Plicatula gibbosa, Heleobia* sp.). *Ostrea equestris* is dominant (45%).

At *Dianova* (*ca.* 4977 ^{14}C years BP) predominate *Ostrea equestris* too (91%).

The other species are *Erodona mactroides* and *Mytilus edulis*.

The shell midden of *Punta Espinillo* (*ca.* 3500–2000 ^{14}C years BP) contains at least seven species (*Heleobia* sp., *Ostrea equestris, Tagelus plebeius,* unidentified gastropods *Mytilus edulis, Buccinanops deformis, Plicatula gibbosa, Siphonaria lessoni*). The dominant species are *Mytilus edulis* (36%), *Heleobia* sp. (35%) and *Ostrea equestris* (23%).

Four species are present in *Gambé* (*ca.* 3000 a 2000 ^{14}C years BP): *Heleobia* sp. *Erodona mactroides, Tagelus plebeius* and *Ostrea equestris*, predominating the first one (98%).

The site *Colonización* 6 (*ca.* 3000 a 2000 ^{14}C years BP) has at least seven species: *Erodona mactroides, Heleobia* sp., *Tagelus plebeius, Anomalocardia brasiliana, Mactra* sp., *Mytilus edulis*, and unidentified gastropods. *Erodona mactroides* and *Heleobia* sp. are the most abundant species, being the 37% and 63% respectively. Only *Erodona mactroides* is present in the *Colonización* 7 and 8 shell middens (*ca.* 3000 a 2000 ^{14}C years BP). At *Km 26* (*ca.* 3000 a 2000 ^{14}C years BP) there are two species: *Erodona mactroides* and *Heleobia* sp., predominating the first one (96%).

When comparing the mollusks composition among all shell middens, the differences are statistically significant (Kruskal-Wallis test Hc = 66.9, p = 0.0001922 < 0.05). As seen in Fig. 6.6, the richest site are *Punta Espinillo*, *Colonización* 6 and *Cañada de las Conchas*. *Punta Espinillo* and *Colonización* 6 have shelly mounds with heights of 1–2 m, and with dimensions of 200 × 50 m. They correspond with the U and S types of shells middens defined by Beovide (2011b). *Cañada de las Conchas* is mostly a mound, but his oldest parts (*ca.* 7000 a 6000 ^{14}C years BP) are really small rubbish dumps.

6.4.1 Habitat of the Mollusks

The bivalves *Erodona mactroides, Tagelus plebeius, Mactra* sp., *Anomalocardia brasiliana* and the gastropods *Heleobia* sp. and *Buccinanops deformis* live in soft substrates in the intertidal or subtidal zones.

Ostrea equestris, Mytilus edulis and *Plicatula gibbosa* are bivalves that live over hard substrates (rock, other shells) in the intertidal or subtidal zones. The gastropod *Siphonaria lessoni* lives in the coastal rocky points that usually limit the Uruguayan beaches.

In general the mollusks come from distances between 0 and 3 km (Table 6.3), immediate supply, and there is only one case with a distance of 3–10 km (*Mytilus edulis* in Dianova).

6.4.2 Correspondence Analysis

The first factor explain 45.1% and the second 24.5% of the total variability, respectively (together 69.6% of the total). The analysis show that the most of the resources are within an area of about 10 km of radius, are potentially dense, predictable, and with an annual disponibility (Table 6.4, Fig. 6.7).

6.5 Discussion

The spatial location pattern of the eight shell middens from the lower part of the *Santa Lucia* river basin is dispersed (not at random). The spatial organization of human groups can be seen as an strategy to take control of the natural resources (Dyson-Hudson and Smith 1983) and the exploitation pattern is by chance or to some conditioning factor, natural or cultural. A hypotheses to explain the dispersed

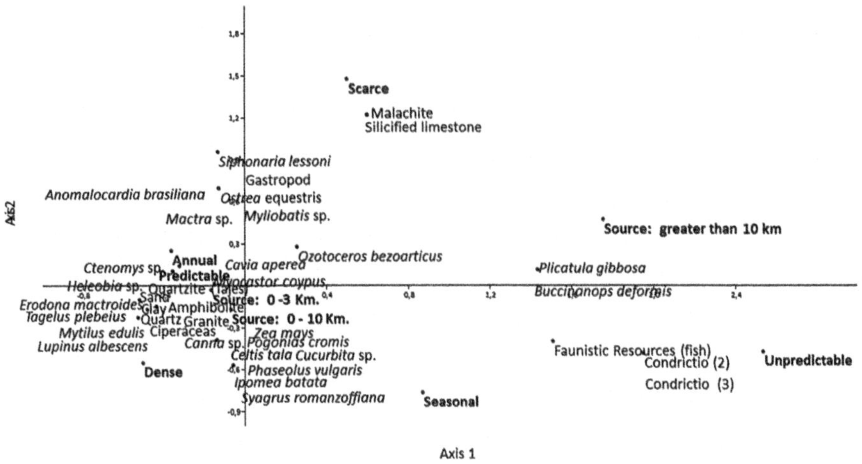

Fig. 6.7 Plots of scores on the first two correspondence analysis axes, showing the variables analyzed (Table 6.4)

distribution, is that the spatial arrangement of the shell middens obey to the use of the resources of the basin.

Species accumulation curves show the rate at which new species are found within a community, and can be used for estimating species richness (Gotelli and Colwell 2010). Curves with rapid accumulation or rarefaction could be associated with environmental heterogeneity and different availability of resources (species). Curves with linear trends or without trend (*Gambé*, Km 26) could be associated with domestic activities in residential bases and formed by discarded mollusk valves (Beovide 2011b). *Santa Lucia* river basin is situated in the intermediated estuarine region of the *Rio de la Plata* coastal zone with higher environmental variability (Nagy et al. 1997). Curves from *Punta Espinillo* and *Cañada de las Conchas*, rises very rapidly in the beginning, as most of the common species are detected early in a site. *Colonización* 6 and *Dianova* would be transitions shell middens sites with less influence of the *Río de la Plata* estuary.

The cratonic rocks ("*Basamento cristalino*") are the substrate of the basin, which was filled with Tertiary and Quaternary sediments, forming terraces of *ca.* +2.5 to *ca.* +20 m amsl. The mollusks that live and lived in sandy sediment are ubiquitous in this area. Then, sounds logical that the radius of exploitation are very near the shell middens. Indeed, the rocky outcrops are located only in the left margin of the river, near its mouth, and the mollusks that live attached to the rocks were exploited only from the near sites, or at a radius of 3–5 km. Other resources were exploited too (Fig. 6.8). At the shell middens with *ca.* 6000–3500 ^{14}C years BP (*Cañada de las Conchas*, *Punta Espinillo* and *Dianova*), lithic instruments were made with amphibolite, quartz, silcretes, or granite. Clay and sand were used to make pottery containers since *ca.* 5000 years ago. In the same deposits are remains of fishes

Fig. 6.8 Shell midden *Colonización* 6. (**a, g**) excavation V, (**b**) malachite ornaments, (**c**) amphibolite tools, (**d**) shark tooth (ornaments), (**e**) phalanx of deer (*Ozotoceros bezoarticus*) with butchering and skinning marks (**f**) Maize cob phytoliths and starch grains (*photo*) were recovered from residues on plant grinding tools and selected archaeological sediments corresponding to the shell middens (*Colonización* 6 and *Gambé, ca.* 3000 years ^{14}C BP)

(burned teeth of *Pogonias cromis*) and some plants (*Canna* sp., *Lupinus albescens*) (Beovide 2009, 2011b).

Between *ca.* 3000 and 2000 years BP: *Colonización* 6, 7, 8, *Gambé* and Km 26. They contain botanical resources that could have been harvested or grown, such as: *Zea mays* (corn), *Ipomea batata* (sweet potato), *Cucurbita* sp. (squash), *Celtis tala* (tala), *Canna* sp. (achira). There are also Cyperaceae and two species of legumes (*Phaseolus vulgaris*, kidney bean) and *Lupinus albescens*, and microscopic coal (Beovide 2011a; Beovide and Campos 2014). These species were recognized by the study of phytoliths and starch in lithic grinding tools, interpreted as related to flour production (Beovide and Campos 2014). There are also remains of burned *Opuntia* sp. (tuna) and of *Syagrus romanzoffiana* (pindó palm) in the shell middens.

Although the remains of animals other than molluscan are scarce, they always have evidence of use for ornaments, manufacturing, or food processing. *Pogonias cromis* (black drum), *Myliobatis* sp. (eagle ray), and three species of shark, are aquatic resources found in shell midden (Beovide 2011b). All have evidence of food processing marks and some of manufacturing marks for ornamentation (e.g. shark teeth, Beovide 2011b). *Ozotoceros bezoarticus* (deer), some rodents as *Myocastor coypus* (nutria), *Cavia aperea* (apereá), *Ctenomys* sp. (tuco-tuco) have marks associated with technology or food (traces of fire, marks for skinning, butchering and marrow cracking, among others) (Beovide 2011b).

Mineral resources such as granite and quartz were utilized for plant-grinding lithic tools (mano and mortar) and net sinkers (Beovide 2011b). Silcretes, amphibolite and quartz were employed for points and lithic instrument (Beovide 2011b, 2013). Green stones (e.g. constituted of malaquite) were used to make ornaments (Beovide 2011b). Sand and clay were utilized to make pottery (Beovide 2011b).

The spatial organization of human groups can be seen as an strategy to take control of the natural resources (Dyson-Hudson and Smith 1983). To analyze the relationship between the human settlements with the resources which are in the sites (Bettinger 1991; Dyson-Hudson and Smith 1983, among others) the potential resource predictability is taken into account as well (predictability has both a spatial component—predictability of location—and a temporal one—predictability in time-) (Dyson-Hudson and Smith 1983). Potential resource density refers to the potential scarcity or abundance—in relation with the environmental dynamics—of the resources recovered from the shell middens. Seasonality potential of the identified resources makes reference to their annual or seasonal cycles.

According to potential distribution taking into account the paleogeography between *ca.* 6000 and 2000 years BP (Beovide 2011b), correspondence analysis result, their predictability (predictable/unpredictable), their potential density (taking into account the paleoenvironment) and their ciclicity (annual/seasonal) most of the resources found in the shell middens they were present within an area of about 10 km of radius, were potentially dense, predictable, and with annual disponibility. Areas with a radius of 3 and 5 km (Fig. 6.9), are adequate to explain the use de mollusks resources and their potential areas of catchment in the sites. This situation corresponds to the statements of (Higgs and Vita-Finzi 1972) about the use of the

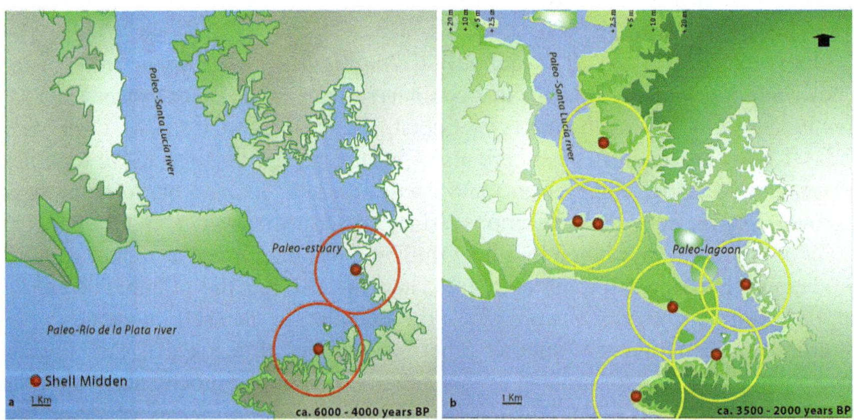

Fig. 6.9 Buffers with radius of 3 km from the shell middens analyzed: (**a**) 7000–4000 years BP, (**b**) *ca.* 3000–2000 years BP

resources in nurseryman societies, with the caution about the limitations perceived by Brooks (1989–1990) in his studies about catchment.

6.6 Conclusions

The location of the shell middens is not at random, and it is determinated by the location of the sites in places near the resources, from *ca.* 7000 years BP.

The majority of the mollusk resources are within an area with a radius of about less than 5 km, and the other resources could be find radius of less than 10 km. According to Dyson-Hudson and Smith (1983) it corresponds with a stable territorial system and with residential sites, with low mobility and high defendability of resources. It seems then that the *Santa Lucia* basin societies had reduced residential mobility and practiced the horticulture, and the spatial pattern of the shell middens points in the same direction. This pattern implies the necessary intersection of the potential catchment provision areas with a 3–5 km of radius, especially in the case of the archeological shell middens of *Gambé* and *Colonización*, which are contemporaneous. This indicates that the people from these sites interacted in the same environment. In consequence, we are dealing with a unique group or closely related groups, or different populations that coexisted in the same territory. These hypotheses will be explored in future works.

Acknowledgments Fieldwork during 2011–2015 was supported by grants from the National Agency for Research and Innovation, project FCE-5818 and DICYT-MEC, Uruguay. We are grateful to our research team for the archaeological work.

References

Benzecri JP (1980) L'Analyse des Donnees. 2 L'Analyse des Correspondences. Dunod, Paris, 632 pp

Beovide L (2001) Recursos y organización del espacio prehistórico costero en la Cuenca Inferior del Rio Santa Lucia, Uruguay. In: Beovide L et al (eds) X Congreso Nacional de Arqueología Uruguaya: La Arqueología Uruguaya Ante Los Desafíos Del Nuevo Siglo. AUA, Montevideo, pp 1–30

Beovide L (2009) Transformaciones productivas y dinámica costera: masalla del concepto de - cazadores-recolectores prehispánicos. In: Laporte J et al (eds) XXIII Simposio de Investigaciones Arqueológicas en Guatemala, vol 1. Asociación TIKAL, Guatemala City, pp 223–236

Beovide L (2010) La presencia humana en el curso medio del Rio de la Plata (Uruguay) durante el Holoceno Medio – reciente: una perspectiva de la continuidad y el cambio. In: Barcena R, Chiavazza H (eds) XVII Congreso Nacional de Arqueología Argentina, Arqueología Argentina en el Bicentenario de la Revolución de Mayo, vol 1. Zeta Editores, Mendoza, pp 333–338

Beovide L (2011a) La presencia de cultigenos desde el quinto milenio en el registro del curso medio Platense. In: Feuillet R et al (eds) Avances y perspectivas en la Arqueología del Nordeste. Santísima Trinidad (ST Ser. Gráficos), Buenos Aires, pp 155–173

Beovide L (2011b) Arqueozoología de los depósitos conchíferos de la cuenca inferior del rio Santa Lucia, Uruguay. Doctoral thesis, Universidad de la República, Montevideo, 450 pp

Beovide L (2013) Concheros en la costa uruguaya del Río de la Plata: una aproximación a la explotación y uso de moluscospor las sociedades de fines del Holoceno Medio. Cuad Inst Nac Antropol Pensam Latinoam, Ser espec 1(1):136–148

Beovide L (2014) Shell middens and the use of molluscs in the Late Middle Holocene in the Rio de la Plata: an ethnoarchaeological contribution. In: Szabo K et al (eds) Archaeomalacology: shells in the archaeological record, British archaeological reports, international series, vol 1. Archaeopress, Oxford, pp 111–121

Beovide L, Campos S (2014) El manejo del entorno vegetal y cultigenos (*Zea mays L.*) tempranos en los concheros entre *ca.* 3000 y 2000 años AP, en la cuenca inferior rio Santa Lucia, Uruguay. In: Beovide L et al (eds) Segundo Congreso Internacional de Arqueología de la Cuenca del Rio de la Plata, Libro de Resúmenes, San José de Mayo, p 113

Beovide L, Martínez S (2014) Concheros Arqueológicos en la Costa Uruguaya: Revisión y Perspectivas. Rev Chilena Antropol 29:26–32

Beovide L, Caporale M, Baeza J (2001) Arqueología costera en el area de la Cuenca Inferior del rio Santa Lucia. In: Beovide L et al (eds) X Congreso Nacional de Arqueología Uruguaya: La arqueología uruguaya ante los desafios del nuevo siglo, L. AUA, Montevideo, pp 1–30

Beovide L, Martínez S, Norbis W (2014a) Discriminación entre acumulaciones de moluscos naturales, antrópicas modernas y arqueológicas, constituidas por las mismas especies. In: Beovide L et al (eds) Segundo Congreso Internacional de Arqueología de la Cuenca del Rio de la Plata, Libro de Resúmenes, San José de Mayo, p 103

Beovide L, Martínez S, Norbis W (2014b) Etnobiología de *Erodona mactroides* (Mollusca, Bivalvia): análisis espacial y tafonómico de concheros actuales. Etnobiologia 12(2):5–20

Bettinger R (1991) Hunter-gatherers: archaeological and evolutionary theory, 1st edn. Springer, New York

Binford L (1980) Willow smoke and dog's tails: hunter-gatherer settlement systems and archaeological site formation. Am Antiq 45:4–20

Binford L (2001) Constructing frames of reference: an analytical method for archaeological theory building using ethnographic and environmental data sets. University of California Press, Oakland

Bracco R (1994) Un fechado de Carbono 14. In: Relevamiento, Diagnostico y Rescate Arqueológico en el área de Punta Espinillo (Departamento de Monte-video). Facultad de Humanidades y Ciencias de la Educación de la Universidad de la Republica - Intendencia Municipal de Montevideo, Montevideo, pp 113–114

Bracco R, Del Puerto L, Inda H, Castiñeira C (2005) Middle-late Holocene cultural and environmental dynamic in the east of Uruguay. Quat Int 132:37–45

Bracco R, Del Puerto L, Inda H, Panario D, Castiñeira C, García-Rodríguez F (2010) The relationship between emergence of mound builders in SE Uruguay and climate change inferred from opal phytolith records. Quat Int 245:62–73

Brooks R (1989–1990) Una evaluación crítica del análisis del "catchment". Etnia 34–35:9–45

Cavallotto J, Violante R, Parker G (2004) Sea-level fluctuations during the last 8600 years in the de la Plata river (Argentina). Quat Int 114:155–165

Clark P, Evans F (1954) Distance to nearest neighbor as a measure of spatial relationships in populations. Ecology 35:445–453

Colwell R, Mao C, Chang J (2004) Interpolating, extrapolating, and comparing incidence-based species accumulation curves. Ecology 85:2717–2727

Dincauze D, Driver J (2003) Environmental archaeology: principles and practice. Can J Archaeol 27(1):121–124

Dyson-Hudson R, Smith E (1983) Territorialidad Humana: Una reconsideración ecológica. Cultura y Ecología. In: Buxo Rey M (ed) Las sociedades Primitivas. Editorial Mitre, Barcelona, pp 151–185

Favier C, Borella F (2007) Consideraciones acerca de los procesos de formación de concheros de la costa norte del Golfo San Matías, Río Negro. Cazadores Recolectores del Cono Sur 2:151–165

Franco N (2004) La organización tecnológica y el uso de escalas espacial es amplias. El caso del sur y oeste de Lago Argentino. In: Acosta et al (eds) Temas de Arqueología: Análisis Lítico. Departamento de Publicaciones, UnLU, Buenos Aires, pp 101–145

Giovas C (2009) The shell game: analytic problems in archaeological mollusc quantification. J Archaeol Sci 36:1557–1564

Gotelli N, Colwell R (2010) Estimating species richness. In: Magurran AE, McGill BJ (eds) Frontiers in measuring biodiversity. Oxford University Press, New York, pp 39–54

Gould R (1978) The anthropology of human residues. Am Anthropol 80:815–835

Greenacre MJ (1984) Theory and applications of correspondence analysis. Academic Press, London, 364 pp

Hammer Ø, Harper D, Ryan P (2001) PAST: Paleontological Statistics Software Package for education and data analysis. Palaeontol Electron 4(1):9. http://palaeo-electronica.org/2001_1/past/issue1_01.htm

Higgs E, Vita-Finzi C (1972) Prehistoric economies: a territorial approach. In: Higgs E (ed) Paper in economic prehistory. Cambridge University Press, Cambridge, pp 27–36

Hodder I, Orton C (1990) Análisis espacial en Arqueología. Editorial Crítica, Barcelona

Iriarte J (2006) Vegetation and climate change since 14,810 14C yr B.P. in south-eastern Uruguay and implications for the rise of early Formative societies. Quat Res 65:20–32

Kotzian C, Simões M (2006) Taphonomic signatures of the recent freshwater mollusks, Touro Passo Stream, RS, Brazil. Rev Bras Paleontol 9:243–260

Krebs C (1999) Ecological methodology, 2nd edn. Benjamin Cummings, Menlo Park, California

Martínez S, Rojas A (2013) Relative sea level during the Holocene in Uruguay. Palaeogeogr Palaeoclimatol Palaeoecol 374:123–131

Martínez S, Rojas A, Ubilla M, Verde M, Peréa D, Piñeiro G (2006) Molluscan assemblages from the marine Holocene of Uruguay: composition, geochronology, and paleoenvironmental signals. Ameghiniana 43(2):385–397

Mason R, Peterson M, Tiffany J (1998) Weighing vs. counting: measurement reliability and the California School of Midden Analysis. Am Antiq 63(2):303–324

Nagy G, Martínez C, Caffera RM, Pedrosa G, Forbes EA, Perdomo AC, López J (1997) The hydrological and climatic setting of the Río de la Plata. In: Wells, Daborn (eds) The Rio de la Plata: an environmental overview. An ECOPLATA project background report. Dalhousie University, Halifax, pp 17–68

Sokal R, Rohlf F (1995) Biometry: the principles and practice of statistics in biological research, 3rd edn. W. H. Freeman, New York

Suárez R, López J (2003) Archaeology of the Pleistocene–Holocene transition in Uruguay: an overview. Quat Int 109–110:65–76

Urien C (1970) Les Rivageset plateau continental du Sud du Bresil, de l'Uruguay et de l'Argentine. Quaternaria 12:57–69

Violante R, Parker G (2004) The post-last glacial maximum transgression in the de la Plata River and adjacent inner continental shelf, Argentina. Quat Int 114:167–181

Zuschin M, Stachowitsch M, Stanton R (2003) Patterns and processes of shell fragmentation in modern and ancient marine environments. Earth Sci Rev 63:33–82

Use of Animals During the Mid-Archaic and the Initial Period in Pernil Alto: A Site in the Palpa Valleys, Southern Coast of Peru

Carmen Rosa Cardoza, Johny Isla, Markus Reindel, Enrique Angulo, Hermann Gorbahn, and Lucía Watson Jiménez

7.1   Introduction

This archaeozoological study refers to the use of faunal resources and their management by humans during two very specific periods: the Mid-Archaic (3600–3000 BC) and the Initial Period (1500–800 BC), both documented at Pernil Alto, a site located in the Palpa valleys, on the southern coast of Peru (Fig. 7.1).

These dates are based on 40 ^{14}C samples collected in Pernil Alto from well-documented archaeological excavations led by Markus Reindel and Johny Isla. All the ages mentioned in the text refer to 1σ of calibration (Unkel et al. 2012; see also Reindel and Isla 2009).

C.R. Cardoza (✉)
International Council for Archaeozoology (ICAZ), Lima, Peru
e-mail: cardozacr@gmail.com

J. Isla
Andes: Centro de Investigación para la Arqueología y el Desarrollo, Lima, Peru
e-mail: isla.nasca@gmail.com

M. Reindel
Deutschland Archaeologist Institut (DAI), Bonn, Germany
e-mail: markus.reindel@dainst.de

E. Angulo
Freelance Veterinary Consultant, Lima, Peru
e-mail: huamachito@yahoo.com

H. Gorbahn
University of Kiel, Kiel, Germany
e-mail: hgorbahn@gshdl.uni-kiel.de

L.W. Jiménez
Universidad Nacional Autónoma de México, Mexico City, Mexico
e-mail: luciawatson111@gmail.com

© Springer International Publishing AG 2017   103
M. Mondini et al. (eds.), *Zooarchaeology in the Neotropics*,
DOI 10.1007/978-3-319-57328-1_7

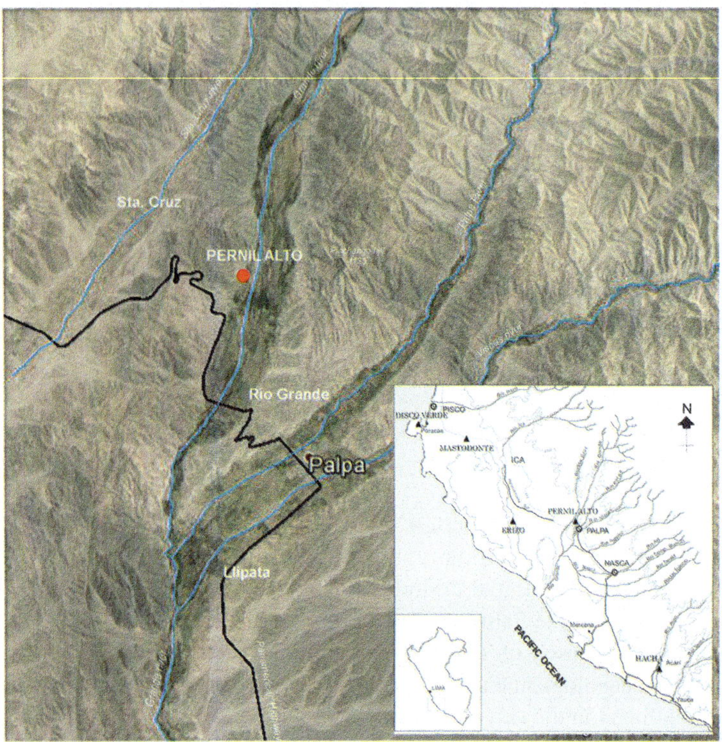

Fig. 7.1 The Palpa valleys on the southern coast of Peru, showing the location of Pernil Alto

Several periods of fieldwork in Pernil Alto, between 2004 and 2009, were carried out with controlled excavations in which an important amount of plant and animal remains were discovered and documented, as well as artifacts and utensils related to obtaining, transforming and producing tools. These indicate the existence—since those early times—of stable and permanent populations in Pernil Alto, which had access to varied resources not only from their immediate surroundings, i.e., the valley, but also from areas farther away such as the Pacific coast and the highlands or the Andean Paramo. Although these are different cultural periods, quite distant in time, the evidence recorded in Pernil Alto allows us to observe, on the one hand, the start of the sedentary process on the southern coast during the Mid-Archaic and, on the other hand, the completion of such processes during the Initial Period, with the establishment of an economy based on agriculture and the full mastery of humans over territory and resources.

Overall, analysis and knowledge of these processes on the southern coast is important in order to understand how the transition occurred between the production and the domestication of plants, the construction of the first houses, the development of mortuary practices, intensive exchange, and finally, the construction of larger settlements which appear to be the result of increased social complexity (Sandweiss

2014). Although the latter is not as well-defined as on the central and northern coast, the evidence from Pernil Alto indicates that the process was completed.

Within this context, this article focuses on the process of settlement in the Palpa valleys through the analysis of animal bone remains belonging to both periods: Mid-Archaic and Initial Period. We analyzed 1361 fragments of animal bones, teeth and antlers. Of this total, 403 fragments belong to the Mid-Archaic and 958 to the Initial Period. In both cases, the taxa were identified and the Minimum Number of Individuals (MNI) was established in such a way that we were able to estimate the use of animals in each cultural period. The comparative analysis of the samples of animal remains from both periods will help us to have a better understanding of the evolution of exchange networks and the use of animals.

7.2   Geographic Location

Pernil Alto is located in the middle area of the Rio Grande Valley, some 6 km north of the town of Palpa, in the Ica Region, on the southern coast of Peru. The site was established on the valley's right bank, occupying the slopes and dry ravines 10 m above the valley border at the foot of low and rocky hills which are part of the high and steep mountains that border this side of the valley (Fig. 7.2).

This part of the valley is characterized by the presence of broad farming lands, almost permanent water availability from the river bed and the existence of important plant coverage along the river banks and the valley itself. Although these conditions do not seem to have changed much since pre-Hispanic times, evidently there was more humidity previously and, hence, larger amounts of

Fig. 7.2 Aerial view of Pernil Alto and the surrounding area with the locations of the two main sectors (PAP-265 and 266) identified in the site

water. Paleoclimatological studies in the Palpa valleys, both on the coast and in the highlands, show that the conditions of greater humidity that characterized the Holocene changed gradually during the Archaic Period and there is some evidence of droughts and the initiation of a desertification process in the region towards the end of the Archaic, conditions that apparently increased substantially towards the second millennium BC, that is, almost at the beginning of the Initial Period (Eitel et al. 2005; Eitel and Mächtle 2009). This situation probably led the population to settle in places closer to the valley bottom and water sources, as is the case of Pernil Alto.

On the other hand, the location of Pernil Alto in an area quite far from the ocean (60 km in a straight line) and just at the foot of the first hills of the western Andes mountain range provides it with special conditions because it is located at an intermediate point between the coast and highlands. This allowed its inhabitants to have access not only to products farmed in the valley, but also to resources from quite distant and different ecologic zones.

7.3   Site Description

The first archeological record of Pernil Alto was obtained in 2000, a time when many archeological prospection studies were undertaken along the Rio Grande, Palpa and Viscas valleys (Reindel and Isla 2006). Since then, the site has been studied in successive years through archeological excavations in the area aimed at discovering and documenting, as completely as possible, the different occupations identified in the site, particularly those related to the Mid-Archaic Period and the Initial Period (Reindel and Isla 2009; Reindel 2009; Isla 2010).

The site has been divided in three sectors according to surface evidences and surface archeological traits (see Fig. 7.2). The first one occupies a smooth slope in the southern area of the site, close to the valley border, where a number of mud constructions corresponding to the Initial Period were found (PAP-266). An important occupation from the Archaic Period has also been recorded in this sector. The second sector occupies a broad elongated ravine in the central part of the site, where ill-preserved remains of a series of stone terraces organized as steps up to the foot of surrounding hills were observed (PAP-265). The main occupation here dates back to the Initial Period, although looted tombs from the Early Paracas period and a Mid-Nasca domestic occupation have also been recorded in this part of the site. Finally, the third sector occupies a small ravine located to the north of this site between two small rocky hills, where the remains of several housing terraces—related to layers with waste from domestic activity (simple pottery, ash remains, etc.)—were found. The occupation, in this case, also corresponds to the Initial Period.

7.4   Excavations

Excavations at Pernil Alto have focused on the southern side of the site, originally called PAP-266. The first excavations were carried out in 2001, but additional and larger excavations were carried out in 2004 and 2005, when most of a small settlement from the Initial Period was uncovered. It was represented by an architectural foundation of low mud walls. At the end of the 2005 excavations, the remains of another, older occupation were identified under the Initial Period constructions; these were excavated during another period of fieldwork between 2006 and 2009. These excavations, done by cutting part of the Initial Period structures, allowed for the discovery of several half-buried rooms of a circular shape bounded by stones, which correspond to a small settlement from the Mid-Archaic Period.

Additionally, excavations were performed in the central part of the site, recorded as PAP-265, where an Initial Period domestic occupation was recorded. It was made up of housing terraces organized as steps following the slope line. Such terraces were built with retention walls made of stones linked to simple floors rammed on artificial fill.

Thus, excavations at Pernil Alto document evidence of two overlapping occupations from different times: the first one from the Mid-Archaic and the second one from the Initial Period. To date, these are two of the oldest occupations recorded in the Palpa valleys (Reindel 2009; Isla 2010). Occupations pertaining to the Paracas and Nasca cultures have also been recorded in Pernil Alto. However, they are of a lesser scale than those from the Archaic and Initial Periods.

7.5   Site Chronology

Archaeological research in Pernil Alto has allowed the recording and documentation of a long occupational history that spans more than 4000 years, from the Mid-Archaic Period (3600 BC) to the Mid-Nasca Period (450 AD), with a hiatus of 1500 years still to be filled in between the Archaic and the Initial Period (Fig. 7.3).

The two more important and longer occupations recorded in this site correspond precisely to the Mid-Archaic (3600–3000 BC) and the Initial Period (1500–850 AD). The dates for each of these occupations were obtained through ^{14}C for 40 samples from well-controlled excavations in Pernil Alto (Unkel et al. 2012).

7.5.1   The Mid-Archaic Period

As previously noted, the Mid-Archaic occupation in Pernil Alto was identified beneath the Initial Period constructions, where a group of semi-subterranean circular houses whose walls were covered with stone slabs were found. Postholes around them indicate that they were covered with roofs of simple branches (Fig. 7.4).

YEARS	PERIODS	CULTURES	EPHOCS	CERAMIC STYLE	SITES
1535 AD 1400 AD	LATE HORIZON	Inka / Ica		Inka / Ica	Tambo de Llipata
1100 AD	LATE INTERMEDIATE PERIOD	Ica		Ica	Ciudadela de Huayurí Pinchango Alto Chillo
850 AD					
640 AD	MIDDLE HORIZON	Wari		Atarco Chakipampa Loro	Huaraco La Máquina Palpa
	EARLY INTERMEDIATE PERIOD	Nasca	Late	Nasca 6, 7	Parasmarca
			Middle	Nasca 4, 5	La Muña
50 AD			Early	Nasca 2, 3	Los Molinos
200 BC	TRANSITIONAL		Proto-Nasca	Nasca 1 Ocucaje 10	El Quemado Estaquería Carapo
	FORMATIVE PERIOD — EARLY HORIZON	Paracas	Late	Ocucaje 8, 9	Cerro Paracas Jauranga
			Middle	Ocucaje 5, 6, 7	Jauranga
840 BC			Early	Ocucaje 3, 4	Pernil Alto Mollake Chico
1500 BC	FORMATIVE PERIOD — INITIAL PERIOD			Puerto Nuevo Disco Verde Hacha	Pernil Alto
3060 BC					
3760 BC	ARCHAIC			No Ceramics	Pernil Alto

Fig. 7.3 Chronological chart showing Pernil Alto in the cultural processes of the southern coast

Inside the houses, the floor had remains of the activity of the first inhabitants who populated the Palpa valleys between 3600 and 3000 BC (Reindel and Isla 2013).

The Mid-Archaic occupation in Pernil Alto corresponds to the start of the sedentism process in the region. There are clear indications of the development of incipient agriculture, where farming of flat beans, sweet potatoes, gourds and

Fig. 7.4 Photo showing the semi-subterranean dwellings of the Archaic Period discovered below the occupation of the Initial Period at Pernil Alto

pumpkins stand out. The presence of millstones shows that grains and other products were ground there. Food and raw materials, even from distant regions, such as shells, camelid bones, deer antlers, and obsidian, as well as some tools used for hunting, show that the inhabitants of Pernil Alto were highly mobile and obtained part of their diet through hunting. When the site was abandoned, a number of burials were deposited inside the houses, revealing the funerary customs of that time.

7.5.2 The Initial Period

Initial Period occupation in Pernil Alto is represented by a well-structured orthogonal architectural compound (Fig. 7.5), where four main construction stages, as well as a fifth remodeling stage before the site was abandoned, were identified.

Constructions are made up of solid mud block walls. Several use layers and fill containing various materials have been recorded. Within these layers, there are subsistence waste remains, artifacts and artisanal products, particularly pottery (Reindel and Isla 2009). In the case of the pottery, some distinctive features can be observed in terms of shape and decoration that precede the development of the famous Paracas culture pottery.

This evidence suggests that a group of inhabitants with a sedentary way of life lived in Pernil Alto and that their primary economic activity was based on agriculture and producing goods. At this time, complete domestication of plants and animals had been achieved, including corn, manioc, flat beans, sweet potatoes,

Fig. 7.5 Panoramic view of the architectonic structures of the Initial Period in Pernil Alto

and achira (Indian shot), as well as guinea pigs and camelids. The intermediate location of the site, close to the foothills of the Andes, permitted them to have access to resources from areas as far apart as the coast and the highlands, where they obtained marine resources and obsidian, respectively, with greater intensity than during the Archaic Period.

7.6 Analysis of Animal Remains

We present the preliminary results on the analyzed sample that was made up of a total of 1361 fragments of animal hard materials, including teeth, bones and antlers. Of this total, 403 fragments belonged to the Mid-Archaic and 958 to the Initial Period. Bones were quite fragmented and much deteriorated. However, a number of families were identified.

7.6.1 Methodology

Comparative osteology was used to identify the bone remains from Pernil Alto. This aimed at identifying taxa and then establishing the Number of Identified Specimens (NISP) and the Minimum Number of Individuals (MNI) (Mengoni Goñalons 2006–2010).

The analysis procedure had two main stages:

- First stage: bone cleaning, resulting in a first impression of sample size, and observation of taphonomic condition. This stage may seem the simplest and most monotonous one, but it is a key factor in the analysis because it enables observation of the condition of each bone. Thus, during this stage, we observe if the bones have fractures, if they were the result of slaughtering animals, a part of the manufacture of an artifact, or simply a fracture due to poor conservation or recovery conditions.
- Second stage: morphological observation, comparison and classification of each bone by means of comparative osteology. Comparative study of the sample was based on the collection of modern skeletons at the Museum of Archeology of the Universidad Nacional Mayor de San Marcos in Lima. This resulted in some limitations because the current comparative sample is incomplete. Other materials were also consulted, such as illustrated manuals (Altamirano Enciso 1983) and photographs (Cardoza et al. n.d.). At this stage, taxonomic identifications were established and other data were recorded. These data included the systematization of the analysis that was aimed at determining taxa and age estimation when possible for artifacts not included in the dietary data.

In addition, and in connection to both stages of the study, all bones were reassembled and recorded in individual and context sheets. Due to bone fragmentation and the generally bad conservation of the sample, we were not able to take osteometric measures, and no complete individuals were found. Such bad conservation was the result of human activities related to different uses of the animals, such as butchering, skinning, and so forth, and unintentional postmortem fractures during excavation, storage, etc. Despite this, since the material was found in known cultural contexts, it provides important information about the use of animals which will be useful for future studies.

The bone fragments were classified following the procedure proposed by Angulo and Cardoza (2009, p. 133), which allows grouping at various taxonomic levels, including Class, Order, Family, Genus and Species.

The Class categories include mammals, birds, reptiles, amphibians and several types of fish. Mammal fragments that were not identified more specifically were grouped into two broader categories of Large Mammal and Lesser Mammal, which include larger and mid/small-sized animals, respectively (Wheeler 1985). The Large Mammal category includes anatomically unidentified fragments. Taking into account the sample context, they are either camelids or cervids. A third group was also recorded: Sea Mammals, represented by just three specimens (tooth fragments), included in the artifacts category.

To estimate the age of individuals, we classified bone fragments according to structure, size and bone ossification points (Wheeler 1985). Thus, we found the following age groups:

- Early: fetuses and newborns.
- Young: individuals whose bones are not fused or which had incomplete fusion lines.
- Adults: individuals whose bones are totally fused.

A fourth group called Early-Young (see Angulo and Cardoza 2009), was used for bones which, due to their shape and size, could not be included in the other groups and which were very useful for the total analysis. For statistical purposes, we have included this category within "Young."

7.6.2 Number of Identified Specimens (NISP) and Taxonomic Representation

The objective of estimating the NISP was mainly to determine which taxa were present in the sample. All of the bone remains are fragmented. According to our observations, most of the faunal remains would have an anthropic origin.

Of all items in the sample, 56% were classified under the category of Family. The following Families were identified in the study sample:

- Camelidae: comprised of four taxa, two of which are domestic: *Lama glama* (llama) and *Vicugna pacos* (alpaca), as well as two wild ones: *Lama guanicoe* (guanaco) and *Vicugna vicugna* (vicuña). Undoubtedly, this is the most represented family in the sample, particularly for the Initial Period.
- Cervidae: there are two main species: *Odocoileus virginianus* (white-tail deer) and *Hippocamelus antisensis* (taruca), which were identified by their antlers. In the first case, it is found in the general use context and, in the second case, it is represented in the artifacts category.
- Canidae: includes dogs and foxes. There is a great likelihood that most of the fragments identified as canids are dogs, but the lack of finer morphological indicators prevents their identification.
- Caviidae: generally, *Cavia* sp. We were not able to identify if these were *Cavia porcellus* (domestic guinea pig) or *Cavia tschudii* (wild guinea pig), but probably the Initial Period sample mainly includes *Cavia porcellus*.
- Psittacidae: the sample is very small—five bone fragments—and corresponds to small birds which very probably belong to an Amazonian parrot genus.

The taxonomic distribution of the NISP in relation to the sample of 1244 bone specimens is shown in Fig. 7.6. Some categories were represented in very small numbers, such as fish and amphibians. There are also 117 indeterminate fragments.

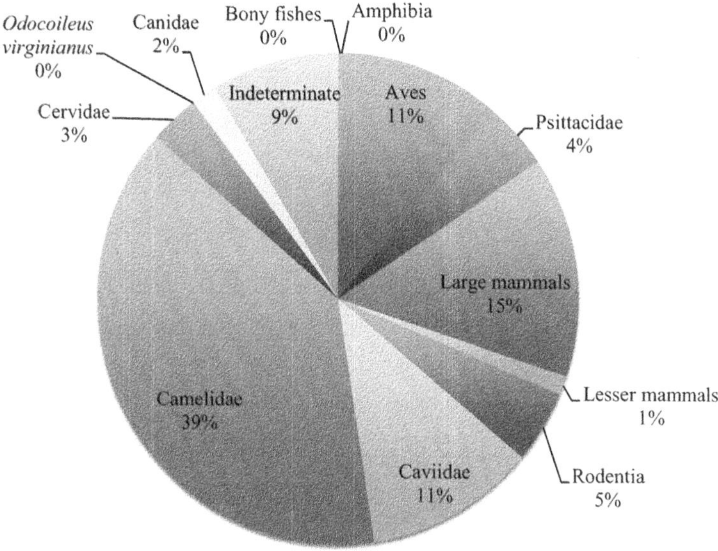

Fig. 7.6 Percentage of Number of Identified Specimens (%NISP) by taxon and indeterminate specimens

This last category is comprised by fragments which could not be placed in any of the preceding categories and is not included in our analysis.

7.6.3   Minimum Number of Individuals (MNI)

The MNI was calculated from the most recurrent bone elements by taxon, laterality, and age category (Mengoni Goñalons 2006–2010). We took into account the concept of "minimum number of individuals with evidence observable in the sample" (Chaix and Meniel 2001). The large and lesser mammals are represented by fragmented bone specimens, while fish, birds and rodents are represented by complete bones.

The MNI was based on the identification of the following bones: vertebrae in bony fishes, long bones in Amphibia, the right humerus in Aves, the left femur in Psittacidae, long bones in large mammals (when morphological indicators were insufficient to determine whether the taxon was camelid or cervid), lumbar vertebrae in lesser mammals, the jaw in Rodentia, Caviidae and Cervidae, the left femur in Camelidae, antlers in *Odocoileus virginianus*, and thoracic vertebrae in Canidae.

In total, MNI = 63 were determined among the 403 fragments belonging to the Mid-Archaic, while MNI = 308 were determined among the 958 fragments from the Initial Period. The two periods altogether include a total of 371 individuals (MNI) (Figs. 7.7 and 7.8), of varied ages (Fig. 7.9).

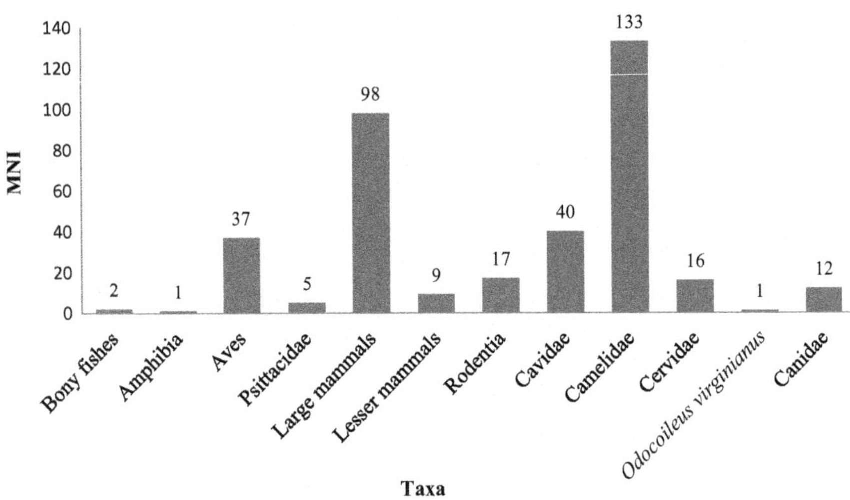

Fig. 7.7 Minimum Number of Individuals (MNI) by taxon in total sample

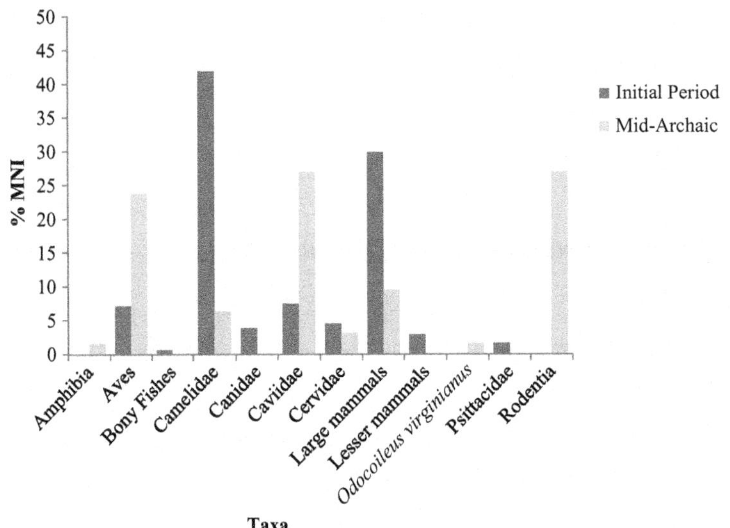

Fig. 7.8 Percentage of Minimum Number of Individuals (%MNI) by period

The sample corresponding to the Mid-Archaic is small and varied. No marked preference is observed for any taxon in particular, although there is an important presence of birds and guinea pigs.

The presence of camelids, mostly mature (Fig. 7.10), is small and would indicate that during that period the use of these animals depended on hunting, as is the case of cervids.

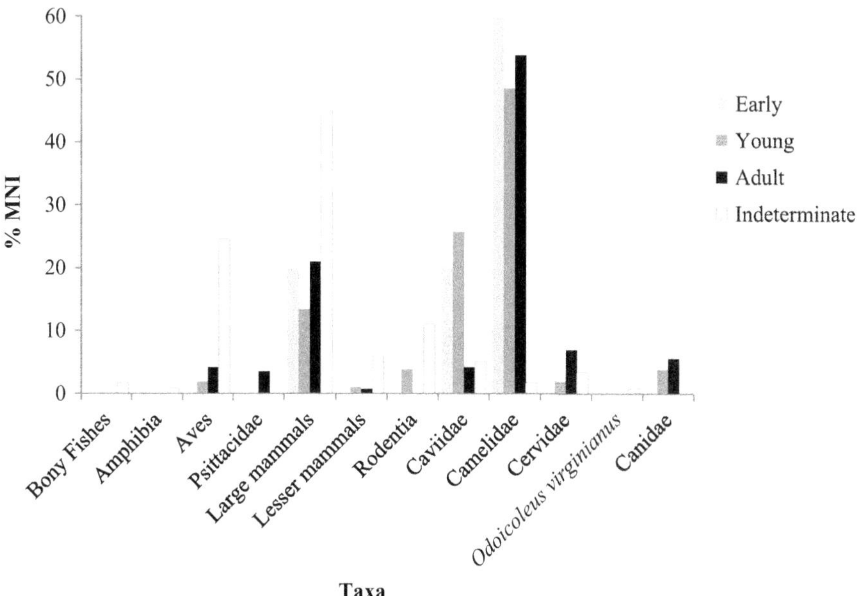

Fig. 7.9 Percentage of Minimum Number of Individuals (%MNI) of the different age groups in total sample

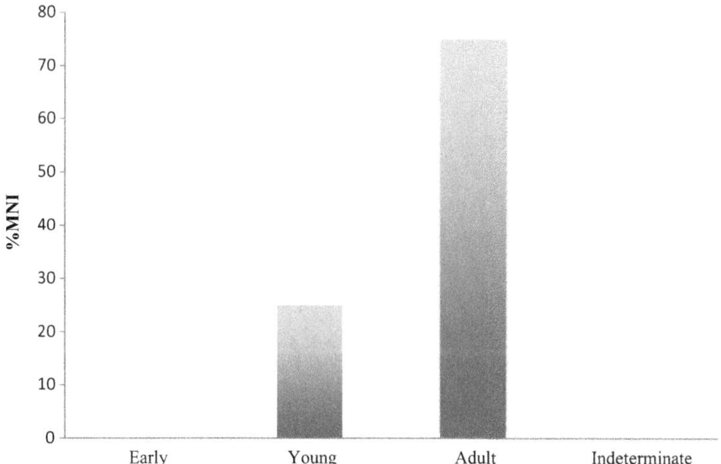

Fig. 7.10 Relative abundance of Camelidae (%MNI) by age group, Mid-Archaic Period

In contrast, during the Initial Period there was a marked preference for the use of camelids, which represents a significant change in human economic activities, implying different habits of protein consumption in these two periods.

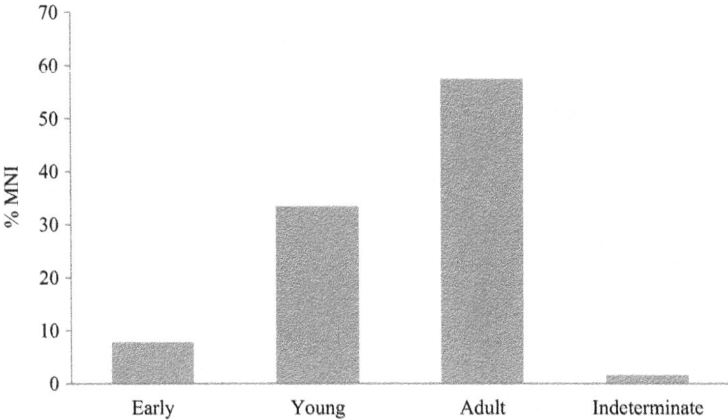

Fig. 7.11 Relative abundance of Camelidae (%MNI) by age group, Initial Period

The study also reveals that during the Initial Period, besides young and adult individuals, there were probably also fetal and newborn camelids (Fig. 7.11), which suggests that at the time, the population managed these animals to a certain extent. This fact becomes more evident if we take into account that there were domestic camelid taxa at the time.

7.6.4   Artifacts

The study sample also includes a significant group of artifacts—made from teeth, bones fragments and an antler—which were studied separately and, therefore, are not counted in the total NISP. These include 28 pieces, all belonging to the Mid-Archaic Period. The classification by Michelle Julien (1985)—who studied the bone industry of the Telarmachay rock shelter in Junín, in the central highlands of Peru—has been taken into account to study these elements.

The analysis focused on the kind of artifact, the element used (Fig. 7.12) and, whenever possible, the taxon (Fig. 7.13).

The analysis revealed that out of all the identified artifacts, 55.6% correspond to the "worked bone" category, which is made up of long bone fragments and ribs with evidence of polishing or wearing, which would indicate the use or production of the piece as an artifact, even if it was not a finished product (which does not necessarily mean it was not used).

Among the most complete artifacts, we identified some ornamental pieces, such as pendants (Fig. 7.14a). In the case of teeth, the pieces were made from the teeth of sea mammals (Fig. 7.14b).

There were also artifacts used for weaving, as is the case of a "needle" made of taruca (*Hippocamelus antisensis*) antler and others to work with soft materials (Fig. 7.15).

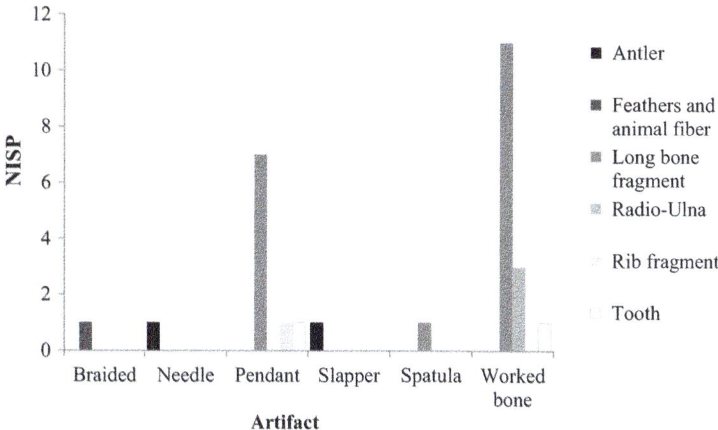

Fig. 7.12 N artifacts (NISP) identified by type made of different animal raw materials

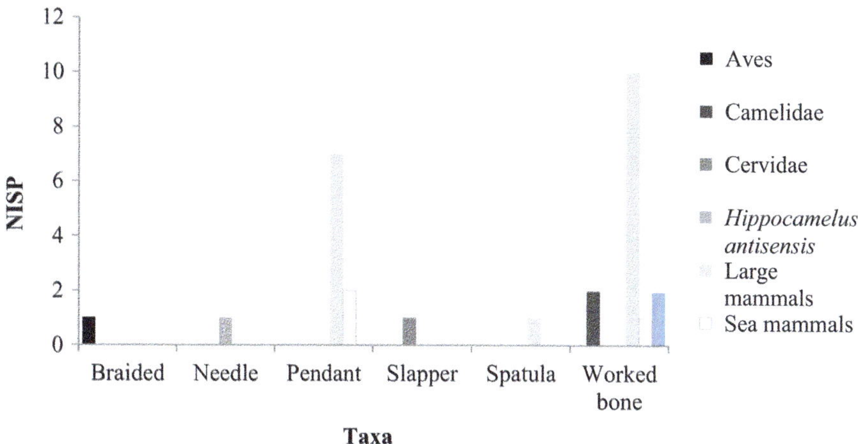

Fig. 7.13 N artifacts (NISP) identified by taxon

An instrument used in the lithic industry was also identified. It is a percussion tool made from a base of shed Cervidae antler, usually used to work stone cores and flake edges. This kind of artifact can also be an early example of a hook for a spear-thrower or atlatl used for hunting, as has been reported by Gorbahn (2013). In the Pernil Alto context, it may have been used to work obsidian and other raw materials.

Fig. 7.14 (**a**) Pendant made from a long bone fragment of a large mammal. (**b**) Pendant made from a sea mammal tooth. (Scale in cm) (Photos Alain Wittmann)

Fig. 7.15 (**a**) Needle made from a *Hippocamelus antisensis* (taruca) antler. (**b**) Artifact made from a metapodial bone of a large mammal. (**c**) Artifact made from the rib of a large mammal. (**d**) Slapper made from a base of molted Cervidae antler. (Scale in cm) (Photos Alain Wittmann)

7.7 Final Comments and Conclusions

The material analyzed here is from a time span that corresponds to an important stage in the process of settling in the Andes and in particular on the southern coast (Engel 1957, 1981; Sandweiss and Rademaker 2013; Sandweiss 2014), where there are few known sites, the archaeological materials of which have not yet been studied in detail. In this sense, Pernil Alto is one of the few settlements systematically excavated which offered the possibility of analyzing and understanding this process, with unique evidence pertaining to the Mid-Archaic Period and the Initial Period (Reindel 2009; Isla 2010).

The Mid-Archaic occupation in Pernil Alto corresponds to the start of the sedentism process in the region, marked by an economy based on horticulture and consumption of wild plants, complemented with marine resources (Gorbahn 2013). During this time, there are clear indications about the development of incipient agriculture. Obsidian points, camelid bones and deer antlers were found, showing that people at this time also had access to far-away resources, such as those located in the highlands. Burials deposited inside the houses reveal the funerary customs of the time. All these features are clearly comparable with those observed in contemporary sites like La Paloma and Chilca, on the central coast (Donnan 1964; Engel 1966, 1980, 1987, 1988; Quilter 1989).

On the other hand, the Initial Period occupation in Pernil Alto is represented by a well-structured architectural compound. Architectural analysis has identified four main construction phases where numerous and diverse subsistence remains, artifacts, and handcrafts—particularly pottery—have been recovered. This evidence shows that the inhabitants of Pernil Alto had a sedentary way of life, whose main economic activity was based on agriculture and production of goods. At this time, a complete domestication of plants and animals had been achieved, while its intermediate location, almost at the foothills of the Andes, also permitted inhabitants access to resources from areas as far as away as the coast and the highlands. In contrast to what was observed in Pernil Alto, previously known evidence for the Initial Period in the region was limited to surface finds and small excavations at sites like Disco Verde in Pisco, Erizo and Mastodonte in Ica, and Hacha in Acarí (Lanning 1960; Rowe 1963; Riddell and Valdez 1987–1988; Engel 1991; Robinson 1994; García and Pinilla 1995).

In this context, some comments and preliminary conclusions can be drawn after analyzing the animal bone remains from Pernil Alto, which must be considered as a first step in the process of understanding the use of animals in the Palpa valleys during the Mid-Archaic and the Initial Period.

As indicated above, the Minimal Number of Individuals (MNI) for both periods included in this study is still limited, particularly for the Mid-Archaic Period. The MNI = 310 for the Initial Period accounts for 83.1% of the analyzed sample, while the MNI = 63 for the Mid-Archaic Period accounts for only 16.9%. This difference is probably due mainly to the size of the excavated area and particularly the density of deposits rather than chronological difference or conservation problems.

On the other hand, a comparative analysis of the distribution of families in both periods shows that during the Mid-Archaic Period, the Caviidae family stood out, accounting for 27.0% of the total. On the contrary, during the Initial Period the most representative family was Camelidae with 41.6%, while Caviidae accounted only for 7.4%. Likewise, distribution by age in the sample shows that 38.3% of the individuals correspond to adults and 29.5% correspond to sub-adults (fetuses, Young and Early). The latter suggests that most probably there were already domestic camelids at the time. Therefore, we suggest that some herd management was practiced. This means seeking pastures for these animals in different ecologic niches that do not require much water and are very resistant in the desert. This would point towards the development of an incipient herding activity.

The Large Mammal category provides us with important information because it groups Camelidae and Cervidae. During the Initial Period, we observe a clear preference for larger animals, which means there was an increase in their use. In the case of Camelidae, it is likely that the fiber, skin, bones and marrow were also used (Pozzi-Escot and Cardoza 1986).

Concerning artifacts, the studied sample shows 28 pieces, all belonging to the Mid-Archaic Period. Of this total, 15 pieces (55.6%) correspond to the "worked bone" category, that is, those with unfinished manufacturing. Among the finished bone artifacts, we identified several ornamental pieces (pendant) and utensils (needles, spatulas, etc.) of excellent fabrication, some of which were used for producing weavings or working stone. In 64.3% of the cases, we found that these artifacts were made with bones belonging to large mammals. Therefore, for now, we cannot draw conclusions on the selection of any particular species for their manufacture.

In sum, based on the results obtained, we can say that between the Mid-Archaic and the Initial Period, there was a significant change in the use of animals, particularly Camelidae and, to a lesser extent, Cervidae and Caviidae. Unfortunately, there is a large gap of almost 1500 years between the two periods. This makes it particularly difficult to reconstruct, for example, the process of camelid use. Nevertheless, it is clear that during the Initial Period there was a greater diversity of taxa, with a remarkable preference for the use of camelids as the main source of protein. This shows that Pernil Alto was occupied by a sedentary—stable and permanent—population with full mastery of the environment and its resources.

Finally, we can say that after the pioneering study of bone remains from Cahuachi, from the Nasca culture (Valdéz 1988), the study of Pernil Alto samples provides an important contribution to our understanding of the use of animals on the southern coast in pre-Hispanic times. The results obtained provide important insights into the earlier occupations of the Ica-Palpa-Nasca valley system, during a time period about which little is known. It is also important because it provides a stratigraphic sequence associated with house structures and spanning a time of important cultural changes, such as increasing sedentism and greater reliance on domesticated animals.

References

Altamirano Enciso AJ (1983) Guía osteológica de cérvidos andinos, Serie de Investigaciones, vol 6. Gabinete de Arqueología, Colegio Real, Universidad Nacional Mayor de San Marcos, Lima

Angulo E, Cardoza CR (2009) Arqueozoología: informe del análisis de los restos óseo-animal del sitio de Cashamarca (Tarma–Junín). In: Calderón Lazo MG (ed) Cashamarca: Su ubicación dentro del proceso histórico del Antiguo Perú. Fondo Editorial Cementos Lima, Lima, pp 132–147

Cardoza CR, Angulo E, Wittmann A (n.d.) Manuales osteológicos I. Familia Camelidae. Unpublished manuscript

Chaix L, Meniel P (2001) Archéozoologie. Les animaux et l'archéologie. Editions Errance, Paris

Donnan CB (1964) An early house in Chilca, Peru. Am Antiq 30:137–144

Eitel B, Mächtle B (2009) Man and environment in the eastern Atacama Desert (Southern Peru): Holocene climate changes and their impact on pre-Columbian cultures. In: Reindel M, Wagner G (eds) New technologies for archaeology: multidisciplinary investigations in Palpa and Nasca, Peru. Springer, New York, pp 17–37

Eitel B, Hecht S, Mächtle B et al (2005) Geoarchaeological evidence from desert loess in the Nazca-Palpa Region, Southern Peru: paleoenviromental changes and their impact on pre-Columbian cultures. Archaeometry 47(1):137–158

Engel FA (1957) Early sites in the Pisco Valley of Peru: Tambo Colorado. Am Antiq 23(1):34–45

Engel FA (1966) Geografía humana prehistórica y agricultura precolombina de la Quebrada de Chilca. Universidad Nacional Agraria La Molina, Lima

Engel FA (1980) Prehistoric Andean ecology. Man, settlement and environment in the Andes, vol. 1: paloma. Distributed by Humanities Press for the Department of Anthropology, Hunter College, City University of New York, New York

Engel FA (1981) Prehistoric Andean ecology. Man, settlement and environment in the Andes, vol. 2: the deep south. Distributed by Humanities Press for the Department of Anthropology, Hunter College, City University of New York, New York

Engel FA (1987) De las begonias al maíz. Vida y producción en el Perú antiguo. Universidad Nacional Agraria La Molina, Lima

Engel FA (1988) Ecología prehistórica andina. El hombre, su establecimiento y el hombre de los Andes. La vida en tierras áridas y semiáridas. Chilca Pueblo I. Implementos de hueso. C.I.Z.A, Lima

Engel FA (1991) Un Desierto en Tiempos Prehispánicos: Río Pisco, Paracas, Río Ica. Centro de Investigación de Zonas Áridas "CIZA", Universidad Nacional Agraria La Molina, Lima

García R, Pinilla J (1995) Aproximación a una secuencia de fases con cerámica temprana de la región de Paracas. J Steward Anthropol Soc 23(1–2):43–81

Gorbahn H (2013) El sitio de Pernil Alto del Arcaico Medio en el sur del Perú: comienzo de horticultura y sedentarismo en condiciones del Holoceno Medio. Diálogo Andino 41:61–82

Isla J (2010) Perspectivas sobre el proceso cultural en los valles de Palpa, costa sur del Perú. In: Valle L (ed) Arqueología y Desarrollo. Experiencias y Posibilidades en el Perú. Ediciones SIAN, Trujillo, Peru, pp 15–52

Julien M (1985) L'industrie osseux. Quatrième Parti, Chapitre II. In: Lavallee D, Julien M, Wheeler J, Karlin C (eds) Telarmachay. Chasseurs Préhistoriques des Andes. Tome XXVIII des Travaux de l'Institut Français d'Etudes Andines, Paris, pp 215–241

Lanning, EP (1960) Chronological and cultural relationships of early pottery styles in ancient Peru. Dissertation, University of California, Berkeley

Mengoni Goñalons GL (2006–2010) Zooarqueología en la práctica: algunos temas metodológicos. XAMA 19–23:83–113. Mendoza, Argentina

Pozzi-Escot D, Cardoza CR (1986) El consumo de camélidos entre el Formativo y Wari, en Ayacucho. Instituto Andino de Estudios Arqueológicos (INDEA)/Universidad Nacional San Cristóbal de Huamanga, Ayacucho, Peru

Quilter J (1989) Life and death at Paloma: society and mortuary practices in a preceramic Peruvian village. University of Iowa Press, Iowa City

Reindel M (2009) Life at the edge of the desert–archaeological reconstruction of the settlement history in the valleys of Palpa, Peru. In: Reindel M, Wagner G (eds) New technologies for archaeology: multidisciplinary investigations in Palpa and Nasca, Peru. Springer, Berlin, pp 439–461

Reindel M, Isla J (2006) Evidencias de culturas tempranas de los valles de Palpa, costa sur del Perú. Boletín de Arqueología 10:237–283. Pontificia Universidad Católica del Perú, Lima

Reindel M, Isla J (2009) El Período Inicial en Pernil Alto, Palpa, costa sur del Perú. Boletín de Arqueología 13:259–288. Pontificia Universidad Católica del Perú, Lima

Reindel M, Isla J (2013) Cambio climático y patrones de asentamiento en la vertiente occidental de los Andes del sur del Perú. Diálogo Andino (41):83–99

Riddell F A, Valdez LM (1987–1988) Hacha y la ocupación temprana del valle de Acarí. Gaceta Arqueológica Andina 16:6–10. Instituto Andino de Estudios Arqueológicos (INDEA), Lima

Robinson RW (1994) Recent excavations at Hacha in the Acarí Valley, Peru. Andean Past 4:9–37

Rowe JH (1963) Urban settlements in ancient Peru. Ñawpa Pacha 1:1–27

Sandweiss DH (2014) Early coastal South America. In: Renfrew C, Bahn P (eds) Cambridge world prehistory. University of Cambridge Press, Cambridge, pp 1058–1074

Sandweiss DH, Rademaker KM (2013) El poblamiento del sur peruano: costa y sierra. Boletín de Arqueología 15:275–294. Pontificia Universidad Católica del Perú, Lima

Unkel I, Reindel M, Gorbahn H et al (2012) A comprehensive numerical chronology for the pre-Columbian cultures of the Palpa valleys, south coast of Peru. J Archaeol Sci 39:2294–2303

Valdéz LM (1988) Los camélidos en la subsistencia Nasca: el caso de Kawachi. Boletín de Lima 57:31–35

Wheeler J (1985) De la Chasses à l'élevage. Deuxième Partie. In: Lavallee D, Julien M, Wheeler J, Karlin C (eds) Telarmachay. Chasseurs Préhistoriques des Andes. Tome XXVIII des Travaux de l'Institut Français d'Etudes Andines, Paris

Taphonomy of Surface Archaeological Bone Assemblages in Coastal Patagonia: A Case Study

A. Sebastián Muñoz

8.1 Introduction

Surface bone assemblages are a key source of information to gain a better understanding of the regional properties of the archaeological record in Patagonian coastal landscapes. Several natural and anthropogenic factors are involved in the present-day dynamics shown by the coastal archaeological record, which usually implies different degrees of information quality and resolution as well as mixing of materials (Favier Dubois and Borella 2007; Caracotche and Ladrón de Guevara 2008; Manzi et al. 2009; Hammond et al. 2013). Trying to understand this dynamics is key to build up a regional picture of the coastal archaeological record, the range of processes affecting it and their consequences, namely the resulting differential preservation and degradation of particular records as well as their integrity and resolution (*sensu* Binford 1981). Several decades of ranching have resulted in a major and constant impact on the Patagonian grasslands by means of losing soil coverage and increasing soil impermeability (Coppa 2004). As a result, soil dynamics is altered by a higher evaporation, leading to a decrease in soil humidity as derived from temperature and wind acting on exposed soils (Coppa 2004). Once soil cover disappears, sedimentary particles respond differently to erosion: some get compacted through trampling; others move away by the action of wind. Bones are sedimentary particles with an organic component and, hence, are strongly affected by changes/variations in soil humidity and exposure (Behrensmeyer 1991). On the other hand, natural deposition of bones can also affect the integrity of bone assemblage, particularly through the input of fresh bones of different size to anthropogenic deposits (Borrero 1989, 1990; L'Heureux and Borrero 2002). It is necessary, then, to understand the way this regional soil erosion process, already in

A.S. Muñoz (✉)
Laboratorio de Zooarqueología y Tafonomía de Zonas Áridas, IDACOR, CONICET, Universidad Nacional de Córdoba, Av. H. Yrigoyen 174, 5000 Córdoba, Argentina
e-mail: smunoz@conicet.gov.ar

© Springer International Publishing AG 2017

M. Mondini et al. (eds.), *Zooarchaeology in the Neotropics*,
DOI 10.1007/978-3-319-57328-1_8

progress, is affecting the nature of the zooarchaeological record and, particularly, its implication in zooarchaeological data needed to understand human-animal relationships in the past. In order to address this point, we have analyzed a surface bone deposit at the Santa Cruz river mouth which was collected in three different consecutive seasons as it became exposed. Our aim is to trace, in the assemblages recovered, the resulting potential taphonomic features that can be sensitive to understand the already mentioned processes and its consequences on the studied bone materials.

The current situation in the southern Santa Cruz river mouth can be described as one where extended deflated areas concentrate archaeological materials. As previously stated, the area is being threatened by several erosional processes, such as marine action, overgrazing and faunal turbation. The former acting on present and past coastal beach ridges, while the second involves trampling by sheep and horse livestock and wild fauna—penguins—passing to coastline from their seasonal nests (Cruz et al. 2008) (Fig. 8.1). The Aeolian erosion, in particular, results both from low precipitation values and average wind action capable of moving sand-sized particles (Ercolano et al. 2013). Although these conditions have characterized the whole twentieth century, Ercolano et al. highlight that during the last decades water imbalance increased probably due to the rise of mean annual temperature by 1 °C. This, in turn, would have increased water imbalance and vegetation instability and, hence, contributed to accelerating, more recently, the exposure of archaeological material (Ercolano et al. 2013).

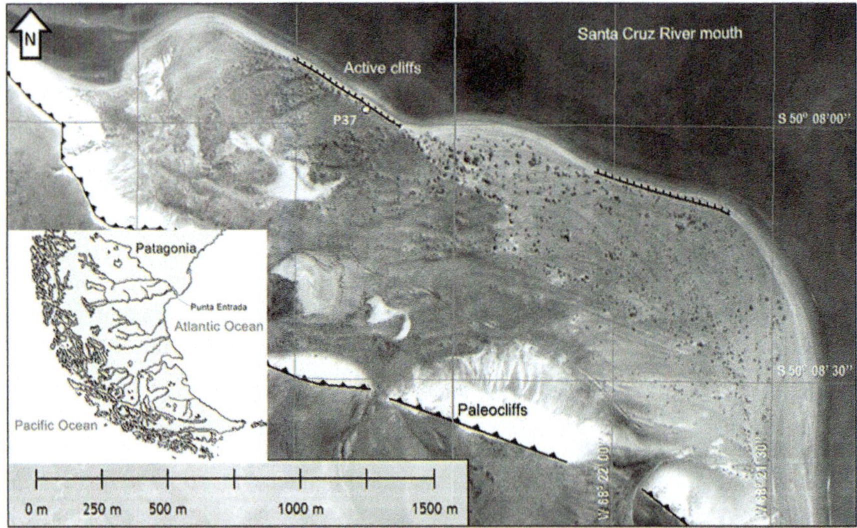

Fig. 8.1 P 37 archaeological site location map

8.2 Objectives, Materials and Methods

In this chapter we analyze zooarchaeological data recovered in three successive collections undertaken at P 37 archaeological site (Fig. 8.2). P 37 is located on the southern bank of Santa Cruz river mouth, in Punta Entrada, a marine accretional landform. This landform is a small territory formed by the sediments that have been transported and deposited by current and by the action of the river during the last 4000 years BP (Fig. 8.1). P 37, in particular, consists of a concentration of archaeological materials that is exposed through erosion. Animal bones are the main materials found in this deposit, followed by lithics and few mollusk shells. The almost complete absence of mollusks makes P 37 an interesting way to explore the reaction of animal bone to exposure conditions, since the matrix is composed of the same sandy particles found on the non archaeological surfaces. Hence, it is expected that archaeological bones do not improve their conservation possibilities by being under more protective conditions such as those provided by shell middens. We discuss the taphonomic signals acquired by bones when they are left exposed and become part of surface assemblages; we consider possible changes in assemblage resolution and taphonomic biases as well as assemblage comparability with other buried material when recovered.

As previously mentioned, P 37 is a concentration of different kinds of archaeological materials, mostly pinniped bones, which lie on a sandy substrate on a 1.5 m height active cliff. The dune where the materials were embedded in was being deflated and the vegetation cover was progressively undermined. An on surface cut marked *Otaria flavescens* bone was radiocarbon dated at 1540 ± 70 years BP—

Fig. 8.2 P 37 archaeological site

1138 ± 70 BP reservoir correction estimation (Muñoz et al. 2009). This dating was confirmed through a second test, measured on a stratigraphic *Arctocephalus australis* bone (LP-3062) indicating 1650 ± 60 years BP, 1250 ± 60 BP after reservoir correction estimation.

P 37 was first localized in March 2006, when a noticeable amount of vertebrate bones were detected on surface and sampled, through 18 collection squares of 2×2 m size over 36 m^2. Sampling of exposed material took place in November 2008, March 2010 and November 2011. At each collection season information on vegetation cover and composition was recorded while all the bone material was picked up for further studies. In order to keep the buried material untouched. The collection was done on bones already exposed, without digging or meshing the sediment. Finally, in November 2011, an additional 1 m^2 stratigraphic sample was taken from C1 square and the fine sediment was screened through a 2 mm mesh. The excavation reached a sterile layer between 18 and 22 cm depth.

In this paper we analyze two 2 m^2 surface collection squares (B2, B3) and the 1 m^2 control stratigraphic sample dug up from C1 square. Thus, a total of four analytical assemblages were isolated. The first assemblage reflects the unknown pre-2008 exposure palimpsest (Assemblage 1: A1). It includes all the material exposed since loss of vegetation cover to the first collection taking place in 2008. Another two assemblages represent bones that were exposed at known time intervals, namely 2008–2010 (A2) and 2010–2011 (A3). Although we cannot be sure that this was the first time they got exposed, we at least know that they were buried at the time of previous collections. Finally, there is a fourth analytical assemblage (A4) which represents the stratigraphic sample from C1 square.

Pinniped bone identification was undertaken with the aid of published osteological guides of *O. flavescens* and *A. australis* (Pérez García 2003, 2008; Sanfelice and Ferigolo 2008). Bone weathering determination follows Behrensmeyer (1978) six-stage sequence for terrestrial vertebrates larger than 5 kg. According to this author, weathering profiles are also presented through three broad categories: fresh (stage 0), slightly weathered (stages 1 and 2) and weathered (stages 3–5). To assess anatomical representation, NISP (number of identified specimens) was calculated and assigned to four anatomical categories of a pinniped skeleton: axial skeleton, front and rear limbs and non determinable limb fragments. Bone modification includes weathering as well as root etching, rodent and carnivore gnawing as well as cut marks (Binford 1981; White 1992). Bone fragmentation considers six size categories as very small (0–1 cm), small (1–2 cm), small-medium (2–4 cm), medium (4–8 cm), large (8–16 cm) and very large (larger than 16 cm).

8.3 Results

We recovered a total of 1028 bones, 875 of which were lying on surface and 153 came from stratigraphy. Most of these remains were identified as pinnipeds, a small part of which could be assigned to species level, namely *A. australis* and *O. flavescens*. These two species, which are osteologically very similar, are well distributed along the Patagonian coast. As shown in Table 8.1, there are also few bird bones and a variable quantity of mammal bones that was not possible to identify at a more specific level.

The assemblage resulting from the first collection (A1) reflects an unknown time of exposure. This is very similar in size to the assemblage recovered later, 16 months ahead, when the site surface was again surveyed to collect a new assemblage (A2). A third assemblage (A3) was recovered in 2011, 20 months after the site surface was cleared away of bones for a second time. Although it represents a longer time interval, it is smaller than the previous ones and similar in size to the stratigraphic assemblage (A4).

Representation of bone fragment size displays a similar trend among these four assemblages, one that is dominated by 2–4 and 4–8 cm bone pieces. Nevertheless some differences are noticeable among them (Table 8.2). The stratigraphic assemblage has a greater proportion of large bone remains (8–16 cm) and a slightly stronger presence of the very small fragments (0–1 cm), something that could be anticipated as the stratigraphic sample included meshing of sediments. In contrast, these categories are underrepresented in A2 and A3, while the 4–8 cm size reaches the higher relative frequency in A1 assemblage as compared to other samples (Table 8.2).

Pinniped anatomical representation is very similar in the four assemblages. Axial bones represent between 45 and 60% of the identifiable bone remains, followed by appendicular bones (rear and front leg) (Fig. 8.3). Undeterminable appendicular bones are more important in A3 and A4 assemblages, something that could be related to the representation of bone size.

Table 8.1 Taxonomical representation at P 37 archaeological site according to the four samples considered in this study

NISP	A1	A2	A3	A4
Pinnipeds	283	311	73	53
A. australis	51	17	5	17
O. flavescens	5	6	1	3
Mammals indet.	31	57	30	79
Avian	2	2	1	1
Total per assemblage	372	393	110	153
Total surface	875			
Total surface + stratigraphy	1028			

Table 8.2 Bone size representation per assemblage from P 37 archaeological site (relative frequencies)

	A1	A2	A3	A4
0–1 cm	3.39	1.16	0.00	5.48
1–2 cm	9.60	19.65	12.35	17.81
2–4 cm	41.53	49.13	46.91	30.14
4–8 cm	35.59	22.83	29.63	17.81
8–16 cm	9.89	6.07	11.11	28.77
> a 16 cm	0	1.16	0	0

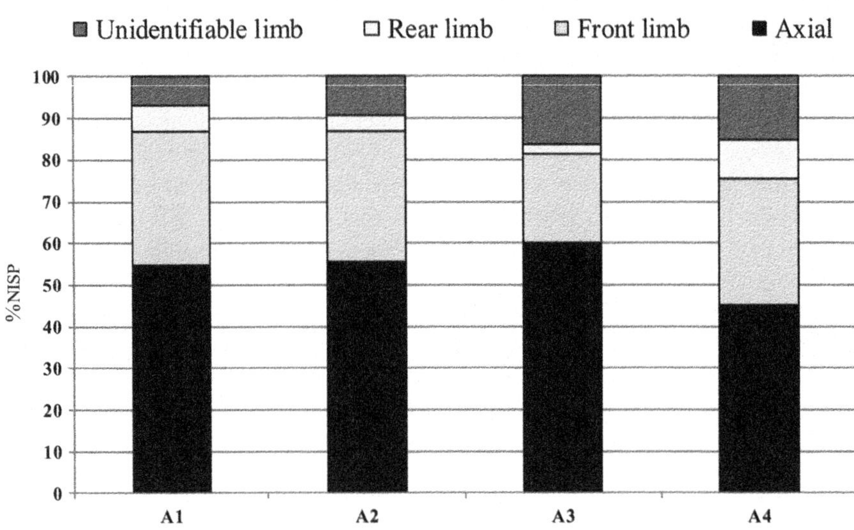

Fig. 8.3 Anatomical pinniped bone representation per assemblage (relative frequencies)

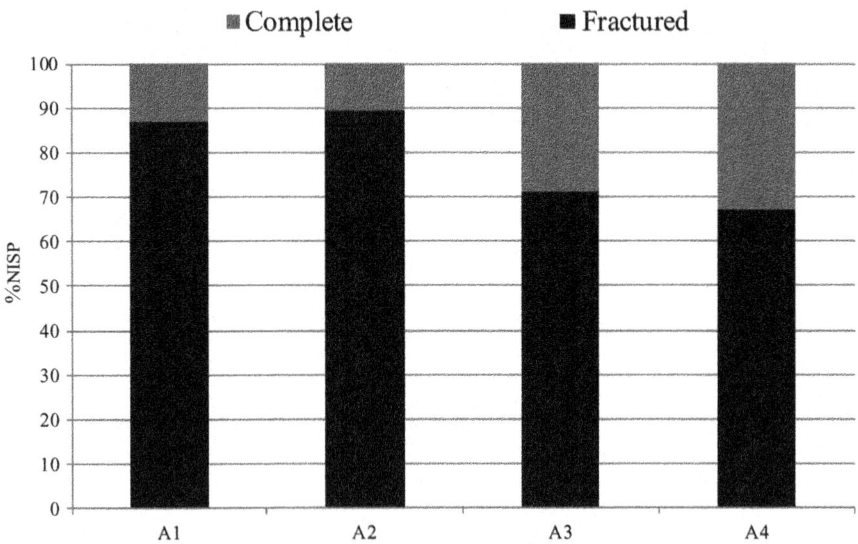

Fig. 8.4 Fragmented bone representation per assemblage (relative frequencies)

Complete bones are better represented in the A4 stratigraphic sample, followed by the A3 surface assemblage, where they represent approximately 30% of the total assemblage in both cases (Fig. 8.4). A1 and A2 assemblages have a higher proportion of broken bones, reaching 90% of the material recovered. Something similar occurs in the representation of unfused bones, consisting of 32–35% of the pinniped specimens recovered in A4 stratigraphic and A3 assemblages and only 15–23% in the first two (A1 and A2 assemblages). Bones which could not be assigned to a fusion state category are, nevertheless, the larger share in the sample exposed (between 41 and 67%) and similar to other categories (27%) in the stratigraphic assemblage.

Two bone modifications differentiate buried from superficial materials, namely bone color and weathering (Fig. 8.5 and Table 8.3). Exposed bones are clearly dominated by a whitish tone that characterizes more than 90% of the material recovered. This color is the result of exposure to solar radiation, a widespread taphonomic signal present on the exposed archaeological bone record of coastal Patagonia (see, for example, Borella et al. 2007; Hammond 2015). A smaller

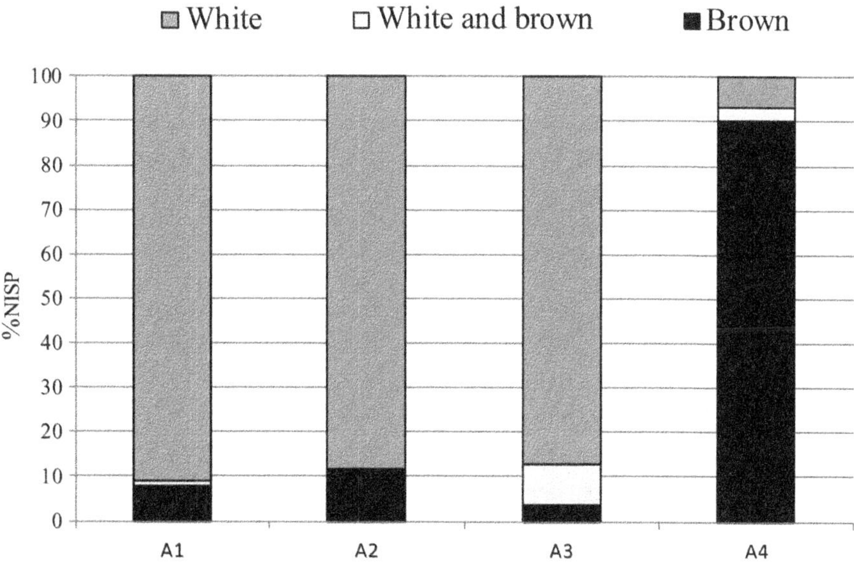

Fig. 8.5 Bone color representation per assemblage (relative frequencies)

Table 8.3 Weathering stage representation per assemblage (relative frequencies)

		A1	A2	A3	A4
Weathering stage representation	0	2.97	1.95	7.79	24.64
	1	34.12	37.01	32.47	63.77
	2	23.74	26.95	29.87	5.80
	3	32.94	31.82	27.27	5.80
	4	6.23	1.95	1.30	0.00
	5	0.00	0.32	1.30	0.00
Bone modifications	Root etching	69.62	54.49	82.28	63.83
	Rodent	1.18	1.80	3.80	5.32
	Cut	7.37	5.99	3.80	11.70

proportion of bones (4–11%) is brown-colored. The opposite happens with buried bones which are 90% brown and only 7% white. Bones displaying a transition between white and brown colors come from A3 assemblage and from stratigraphy (Fig. 8.5).

As a lightly weathered profile dominates all samples, since more than 50% of bone remains can be included in this category (Fig. 8.6). In addition, Fig. 8.6 shows that samples which have been recorded on the surface have a higher proportion of weathered specimens, while fresh bones are more important in A4 sample. The "weathered" category includes 30–40% of bones recovered on surface assemblages but only 6% of those come from stratigraphy. The excavated sample, instead, includes a 25% of fresh bones. This category is underrepresented in the first two surface assemblages, which have less than 3% of bones in fresh condition, while the A3 assemblage rises to 8% of bones with weathering stage 0 (fresh).

Weathering is regularly distributed on bone surface and less than 20% of the specimens display clear indication of contrasting weathering conditions, such as stage 1 and 3 on the same bone and on different surfaces.

When each weathering stage is considered individually it is noticeable that stage 1 dominates exposed and buried bone assemblages. In A4 assemblage, this stage is followed by stage 0, while stage 2 only represents 6% in this assemblage. Instead, surface assemblages have a stage 2 representation between 24 and 30%, similar to that of stage 3. Stage 4 is only present in these assemblages and absent in A4.

Root marks are well represented in the four assemblages, affecting from 54 to 82% of the bone remains, indicating that the deposit had a more stable condition in the past as the result of a grass cover developed on it. Rodent gnawing marks, on the other hand, are scarcely present at P 37 bone materials, although they are relatively

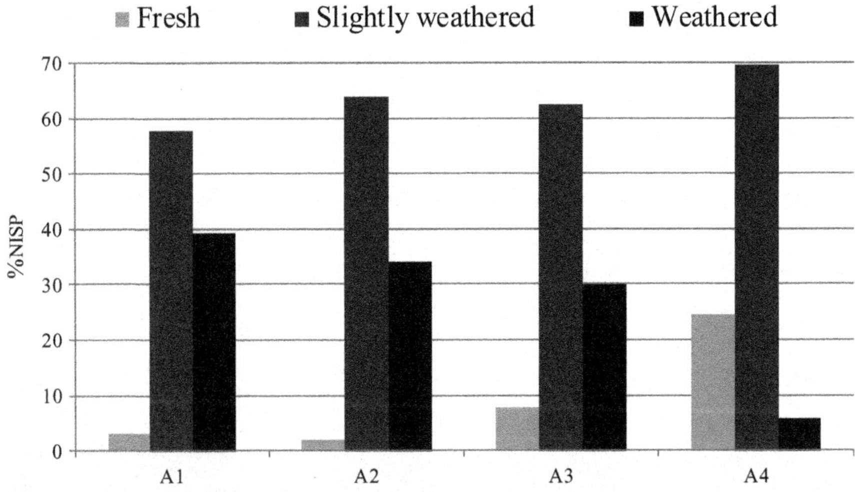

Fig. 8.6 P 37 Bone weathering profiles

more important in A3 and A4 assemblages (Table 8.3). We found no carnivore traces on these bones. Finally, cut marks are better represented among the stratigraphic materials (12%) although they are also important among the A1 materials (Table 8.3).

In sum, taxonomic diversity is similar in all assemblages, although unidentified mammal bone specimens are more important in A4, where very small fragments—usually more difficult to identify—have a better representation. Complete bone specimens are also better represented in this assemblage and also in assemblage A3. In addition, these two assemblages show a higher proportion of bones with fusion state data. A1 and A2, instead, display higher proportions of undetermined fusion state remains.

All assemblages display a slightly weathered profile, although surface bone assemblages also include an important amount of weathered bones. As expected, A4 stratigraphic assemblage includes a higher proportion of fresh bones. All assemblages are also similar in terms of bone modification and anatomical representation. Root etching marks are present in most of the bones recovered while rodent gnawing and cut marks have a slightly better representation in A3 and A4 assemblages. Axial bone fragments represent around half of the identified specimens followed by bone fragments from the rear leg and the front leg.

8.4 Discussion

The current unstable condition of P 37 derives from wind erosion as well as from the action of sea on coastal beach ridges. The site is located on the estuarine high tide line and is affected by the action of water on coastal beach ridges and dunes. This is noticeable in dissected profiles found along the coast.

When P 37 was first located in 2006, grass cover was already lost and most of the bones were on surface. Bones recovered from A1 assemblage displayed signals of sun burning and a slightly weathered/weathered profile (Fig. 8.6). Naturalistic taphonomic studies that are currently undertaken in the area show that developing such a weathering profile in pinniped assemblages requires a minimum of 7 years of exposure (Cruz and Muñoz 2014). Hence, it can be argued that P 37 bone assemblage has an exposure history which exceeds the observation intervals considered during the recovery procedure (16 and 20 months, respectively). The assemblage could have been exposed before final burial and exposition or, alternatively, exposed several times since first disposal. In any case we can conclude that 16/20 month recovery intervals did not differentiate the general taphonomic picture of the exposed assemblage.

There are at least two taphonomic properties that can be linked to the stability phase referred to: the ubiquitous distribution of root etching marks in the entire assemblage and rodent tooth marks. The latter are present in 5% or less in bone specimens from A3 and A4 assemblages and in a lesser proportion in A1 and A2 (Table 8.3). Even when this is not a high value, it can be considered higher than what is usually found in other coastal assemblages such as shell middens (see, for example, Muñoz 2014). Tucu-tuco (*Ctenomys magellanicus*), the most common Patagonian burial rodent, usually chooses grassy dry sandy soils like those found in coastal Patagonia (Nowak 1991). They usually dig a main tunnel 30 cm below surface (Barlow 1969 and Packard 1967 in Nowak 1991) and, as seen in other fossorial rodents, wear away archaeological materials bigger than 6 cm in size while take smaller particles outside the deposit (Bocek 1986). In P 37 rodent tooth mark data are in agreement with this expectation, since 89% of the specimens displaying rodent marks are larger than 5 cm. It can be said, then, that rodents were involved in P 37 taphonomic history and that probably occurred at the time when the site was a stabilized buried deposit covered with a thicker layer of sediments than that what we can see nowadays. There are no signals of galleries or other spatial data that we could use to evaluate their impact on bone assemblage. At present there is not enough sand substrate to allow tuco-tucos to establish there, nor is this sediment compacted enough as to facilitate these actions. The only taphonomic signals left by this process are the tooth marks above mentioned; yet, we do not know whether differences in rodent tooth mark proportions among P 37 samples are the result of weathering bone conditions displayed by surface and stratigraphic assemblages or of a past stratigraphic position that those bones had at the time when the deposit was part of a thicker and stable soil matrix.

When other bone properties are considered, we have found that buried bones show a better representation of complete bones, bone fragments of a of larger size, and there are a higher proportion of bones displaying weathering stages 0 and 1 as well as a higher proportion of unfused bones. Nevertheless, they also share several diagnostic features, such as taxonomic composition, anatomical part representation and some bone modifications. In other words, the buried sample displays a better preservation condition, although not remarkably better. This may be related to the matrix they are embedded in, a sandy substrate which creates an oxidant environment affecting collagen degradation and, hence, bone resistance to physical and chemical agents (Borella et al. 2007).

It is noticeable that the differences observed in taphonomic data between buried and superficial assemblages did not prevent the identification of cut and processing marks (Table 8.3), allowing the study of carcass processing decisions (see Cañete Mastrángelo and Muñoz 2015). This implies that even when bone assemblages display a slightly weathered/weathered profile, such kind of analysis can be approached. In this way, behavioral data coming from deposits other than shell middens can be gathered and studied and a broader spectrum of animal processing decisions can be addressed.

Finally, P 37 can be interpreted as a single depositional unit. We were not able to find clear evidences of different accumulation events such as those informed for other non-shell midden surface bone assemblages located in Punta Entrada (Cruz et al. 2015). For instance, Cruz et al. informed that P 96 surface assemblage, dated between 900 and 1750 years BP, can be interpreted as a palimpsest composed of different taxa and different temporal resolutions in terms of bone deposition and preservation. P 96 palimpsest displays an alternation of exposure and burial episodes which would have affected taxonomical representation. As previously mentioned, P 37 surface and buried assemblages are quite similar and radiocarbon dating standard deviation overlaps each other. The comparative, more homogeneous picture offered by P 37 represents a different part of the range of variability displayed by surface non-shell middens, archaeological assemblages in this sector of coastal Patagonia. Hence, P 37 and P 96 display complementary pictures derived from the incremental erosional activity referred to by Ercolano et al. (2013), nowadays affecting most of the archaeological coastal record of Patagonia (see Cruz and Caracotche 2008).

8.5 Conclusion

Coastal areas have drawn the attention of human populations for a long time, resulting in a variety of coastal archaeological sites and landscape transformations (Anderson 1988; Bailey 1975; Beaton 1985; Erlandson and Moss 2001). The range of variability displayed by geological environments and energy flow is also important, hence, it can be said that change is what is to be expected in this kind of settings (Rapp and Hill 1998; Golberg and Macphail 2006). Coastal Patagonia is not an exception to this general picture, as different research from North to Southern Patagonian coasts clearly shows (Borella et al. 2007; Orquera and Piana 1999). As in other regions of the world, shell middens are one of the most conspicuous anthropogenetic deposits in Patagonian coastal settings (Orquera and Piana 2000; Wells 2001) and, hence, the best known archaeological sites (Favier Dubois and Borella 2007).

Coastal archaeological deposits not embedded in a shell matrix are less known from a taphonomical point of view. It can be asked, then, what assemblage properties and taphonomic signals can be informative of the way buried assemblages, such as P 37, become part of surface archaeological distributions; and how this kind of deposits may inform about the dynamics of the coastal archaeological record in Patagonia.

As we could see, even when it was not possible to discriminate at P 37 changes in most taphonomic variables in time scales shorter than the referred 7-year interval estimation based on actualistic observations, we could identify some taphonomic properties which could be indicative of the transition between burial and exposure conditions in shorter time intervals than those displayed by the general weathering picture of the assemblage (bone color, size and fragmentation). Some other bone modifications, such as rodent and cut marks, are better represented in the stratigraphic assemblage; yet, they are also clearly present in the more weathered assemblages, implying that such lines of evidence could be analyzed even when the general condition of the assemblage is not good. This has already been suggested, recognizing the worth of surface assemblages towards better discussion of human animal interaction in the past (see Muñoz et al. 2013; Muñoz 2015; Borella 2016).

Acknowledgements Carolina Mosconi did the language revision of the manuscript. Figure 8.1 credits correspond to Betina Ercolano. Franca Muñoz helped with final edition of Fig. 8.1. Puerto Santa Cruz Town Council offered logistic support during fieldwork. We thank Víctor López from Estancia Monte Entrance. Carolina Moreno, Patricia Lobbia, Adriana Pretto, Daniela Cañete Mastrángelo and Rubén for field work assistance. Mariela Arriagada, Aldana Calderón and Belén Cippitelli assisted in laboratory analysis. This research has been designed and developed in collaboration with Isabel Cruz and Soledad Caracotche, and was funded by CONICET-PIP-112 201201 00359 CO, SeCyT-UNC PID Res. 313/16 and UNPA 29A/364.

References

Anderson AJ (1988) Coastal subsistence economies in prehistoric southern New Zealand. In: Bailey G, Parkington J (eds) The archaeology of prehistoric coastlines. Cambridge University Press, Cambridge, pp 93–101

Bailey GN (1975) The role of shellfish in coastal economies: the results of midden studies in Australia. J Arch Sci 2:45–62

Beaton JM (1985) Evidence for a coastal occupation time-lag at Princess Charolotte Bay (North Queensland) and implications for coastal colonisation and population growth theories for Aboriginal Australia. Arch Oceania 20(1):1–20

Behrensmeyer AK (1978) Taphonomic and ecologic information from bone weathering. Paleobiology 4:130–162

Behrensmeyer AK (1991) Terrestrial vertebrate accumulations. In: Allison PA, Briggs DEG (eds) Taphonomy: releasing the data locked in the Fossil Record. Plenum Press, New York, pp 291–335

Binford LR (1981) Bones. Ancient men and modern myths. Academic Press, New York

Bocek B (1986) Rodent ecology and burrowing behavior: predicted effects on archaeological site formation. Am Antiq 51(3):589–602

Borella F (2016) Antes del Faro. La explotación de mamíferos marinos en la localidad de arqueológica Faro San Matías durante el Holoceno tardío (Nordpatagonia, Argentina). In: Mena F (ed) Arqueología de Patagonia: De Mar a Mar. Ediciones CIEP/Ñire Negro Ediciones, Coyhaique, pp 295–304

Borella F, Mariano MS, Favier-Dubois CM (2007) Procesos tafonómicos en restos humanos de superficie en la localidad arqueológica de Bajo de la Quinta, Golfo San Matías (Río Negro, Argentina). In: Morello F, Martinic M, Prieto A et al (eds) Arqueología de Fuego-Patagonia. Levantando piedras, desenterrando huesos… y develando arcanos. Ediciones CEQUA, Punta Arenas, pp 403–410

Borrero LA (1989) Sites in action: the meaning of guanaco bones in Fueguian archaeological sites. Archaeozoologia 3(1, 2):9–24

Borrero LA (1990) Taphonomy of guanaco bones in Tierra del Fuego. Quat Res 34:361–371

Cañete Mastrángelo DS, Muñoz AS (2015) El procesamiento de pinnípedos en P37, desembocadura del río Santa Cruz, Patagonia Meridional. Cuad INAPL 24(1):134–152

Caracotche MS, Ladrón de Guevara B (2008) El registro arqueológico costero de Patagonia: diagnóstico del estado actual y herramientas para la conservación. In: Cruz I, Caracotche MS (eds) Arqueología de la costa patagónica. Perspectivas para la conservación. Universidad Nacional de la Patagonia Austral - Subsecretaría de Cultura de la Provincia de Santa Cruz, Río Gallegos, pp 17–45

Coppa RA (2004) El deterioro del pastizal patagónico. Medio Ambiente 4:19–22

Cruz I, Caracotche MS (eds) (2008) Arqueología de la costa patagónica. Perspectivas para la conservación. Universidad Nacional de la Patagonia Austral - Subsecretaría de Cultura de la Provincia de Santa Cruz, Río Gallegos

Cruz I, Muñoz AS (2014) Mammal bone weathering in a temperate coastal steppe (southern Patagonia, Argentina). A comparison among taxa. In: Abstracts 12th conference of the International Council for Archaeozoology (ICAZ), Universidad Nacional de Córdoba, p 43

Cruz I, Muñoz AS, Caracotche MS (2008) Investigaciones arqueológicas y bioantropológicas en la costa atlántica de Patagonia Meridional. In: Libro de Resúmenes, Séptimas Jornadas de Arqueología de la Patagonia. CADIC, Ushuaia, p 48

Cruz I, Ercolano B, Cañete Mastrángelo DS, Caracotche MS, Lemaire CR (2015) Tafonomía y procesos de formación en Punto 96 (Punta Entrada, Santa Cruz, Argentina). Cuad INAPL 24 (1):95–114

Ercolano B, Cruz I, Marderwald G (2013) Registro arqueológico y procesos de formación en Punta Entrada (Santa Cruz, Patagonia Argentina). In: Bárcena JR, Martín SE (eds) Arqueología Argentina, en el Bicentenario de la Asamblea General Constituyente del año 1813. Universidad Nacional de La Rioja, La Rioja, p 527

Erlandson JM, Moss ML (2001) Shellfish feeders, carrion eaters, and the archaeology of aquatic adaptations. Am Antiq 66:413–432

Favier Dubois CM, Borella F (2007) Consideraciones acerca de los procesos de formación de concheros en la costa Norte del golfo San Matías (Río Negro, Argentina). Caz Rec Cono Sur 2:151–165

Golberg P, Macphail RI (2006) Practical and theoretical geoarchaeology. Blackwell, Malden

Hammond H (2015) Sitios concheros en la costa norte de Santa Cruz: su estructura arqueológica y variabilidad espacial en cazadores recolectores patagónicos Tesis para optar al título de Doctora en Ciencias Naturales. UNLP

Hammond H, Zubimendi MA, Zilio L (2013) Composición de concheros y uso del espacio: aproximaciones al paisaje arqueológico costero en Punta Medanosa. An Arqueo Rosario 5:67–84

L'Heureux GL, Borrero LA (2002) Pautas para el reconocimiento de conjuntos óseos antrópicos y no antrópicos de guanaco en Patagonia. Intersec Antropol 3:29–40

Manzi L, Favier Dubois CM, Borella F (2009) Identificación de agentes perturbadores y estrategias tendientes a la conservación del patrimonio arqueológico en la costa del Golfo de San Matías, Provincia de Río Negro. Intersec Antropol 10:3–16

Muñoz AS (2014) La explotación de lobos marinos por cazadores recolectores terrestres de Tierra del Fuego. In: Oría J, Tívoli AM (eds) Cazadores de Mar y Tierra. Estudios Recientes en Arqueología Fueguina. Editora Cultural Tierra del Fuego y Museo del Fin del Mundo, Ushuaia, pp 197–217

Muñoz AS (2015) El registro zooarqueológico del Parque Nacional Monte León (Santa Cruz, Argentina): una perspectiva desde el sitio Cabeza de León 1. Arqueología 21:261–276

Muñoz AS, Caracotche MS, Cruz I (2009) Cronología de la costa al sur del río Santa Cruz: nuevas dataciones radiocarbónicas en Punta Entrada y Parque Nacional Monte León (Provincia de Santa Cruz). Magallania 37(1):39–43

Muñoz AS, Cruz I, Lemaire CR et al (2013) Los restos arqueológicos de pinnípedos de la desembocadura Del río Santa Cruz (Punta Entrada, costa atlántica de Patagonia) en perspectiva regional. In: Zangrando AF, Barberena R, Gil A et al (Comps) Tendencias Teórico Metodológicas y Casos de Estudio en La Arqueología Patagónica. Museo de Historia Natural de San Rafael, Sociedad Argentina de Antropología e Instituto Nacional de Antropología y Pensamiento Latinoamericano, Buenos Aires, pp 459–467

Nowak RM (1991) Walker's mammals of the world, vol 2, 5th edn. John Hopkins University Press, Baltimore

Orquera LA, Piana EL (1999) Arqueología de la Región del Canal Beagle (Tierra del Fuego, República Argentina). Sociedad Argentina de Antropología, Buenos Aires

Orquera LA, Piana EL (2000) Composición de conchales de la costa del Canal Beagle (Tierra del Fuego, República Argentina). Primera parte. Rel Soc Arg Ant 25:249–274

Pérez García MI (2003) Osteología comparada del esqueleto poscráneo de dos géneros de Otariidae del Uruguay. Bol Soc Zool Uruguay, 2ª Época 14:1–16

Pérez García MI (2008) Ontogenia del postcráneo de *Arctocephalus australis* (MAMMALIA, Otaridae). Bol Soc Zool Uruguay, 2ª Época 17:1–19

Rapp G, Hill CL (1998) Geoarchaeology: the earth-science approach to archaeological interpretation. Yale University Press, New Haven and London

Sanfelice D, Ferigolo J (2008) Estudo comparativo entre os sincrânios de Otariabyronia e Arctocephalusaustralis (Pinnipedia, Otariidae). Iheringia, Sér Zool 98:5–16

Wells LE (2001) Archaeological sediments in coastal environments. In: Stein JK, Farrand WR (eds) Sediments in archaeological context. University of Utah Press, Salt Lake City, pp 149–182

White TD (1992) Prehistoric cannibalism at Mancos 5MTUMR-2346. Princeton University Press, Princeton

Luis Manuel del Papa, Luciano De Santis, and José Togo

9.1   Introduction

The activity of burrowing animals in the archaeological record has been a topic of discussion since the beginning of taphonomy. Their importance lies in the fact that fossorial animals can cause a space distortion of the archaeological record when they make their burrows. They can place cultural and natural remains from nearby locations, destroy remains and even incorporate individuals of these fossorial species through natural death in their burrows (e.g. disease, old age, starvation, burrow collapse) (e.g., Bocek 1986; Fowler et al. 2004; Lyman 1994; Mello Araujo and Marcelino 2003; Morlan 1994; Stahl 1996; Weissbrod and Zaidner 2014; Wood and Johnson 1978). Particularly for Argentinean zooarchaeology, in spite of Politis and Madrid pioneer contribution (1988), it has not had a prominent role in the beginning of zooarchaeological analysis. This trend has been reversed throughout the years, since a greater preponderance of studies on rodents are developed to interpret if they were accumulated and deposited by anthropic consumption or natural cause. Most of these works focused on Caviomorpha rodents such as *Ctenomys* spp. (tuco-tuco), *Lagostomus maximus* (vizcacha), *Dolichotis patagonum* (mara or Patagonian hare), *Dolichotis salinicola* (Chacoan mara) and species from subfamily Caviinae (cavies) (e.g., Acosta and Pafundi 2005; Bond et al. 1981; del Papa et al. 2010; Medina et al. 2011; Pardiñas et al. 2011; Quintana and Mazzanti 2011; Salemme et al. 2012; Santiago 2004; Santini 2012; among many others). In

L.M. del Papa (✉) • L. De Santis
Cátedra de Anatomía Comparada, Facultad de Ciencias Naturales y Museo, UNLP, CONICET, Calle 64 entre diag. 113 y 120 s/n, La Plata, Argentina
e-mail: loesdelpapa@hotmail.com; desantis@fcnym.unlp.edu.ar

J. Togo
Facultad de Humanidades, Ciencias Sociales y de la Salud (UNSE), Avenida Belgrano (s) N° 2180, Santiago del Estero, Argentina
e-mail: togofami@arnet.com.ar

© Springer International Publishing AG 2017

137

M. Mondini et al. (eds.), *Zooarchaeology in the Neotropics*,
DOI 10.1007/978-3-319-57328-1_9

recent years, the role of Dasypodidae has also been analyzed at archaeological sites (e.g., Frontini and Deschamps 2007; Frontini and Escosteguy 2011; Salemme et al. 2012; Soibelzon et al. 2013) and to a lesser extent also reptiles (e.g., Albino 1999, 2001; Albino and Franco 2011; Albino and Kligmann 2007; del Papa 2015; Quintana et al. 2002, 2004).

Moreover, throughout the research history in the province of Santiago del Estero, zooarchaeological analyses were scarce and practically neglected despite the pioneering contribution to Argentina of Cione et al. (1979). It is only in the last decade that zooarchaeological analyses of the region became more relevant (del Papa 2012). However, these studies were performed on collections recovered prior to the conception of a taphonomic focus. In this sense, no variables were surveyed during excavation that could be important to differentiate between anthropic and natural input to the archaeological assemblage.

In this paper we present results of the study of fossorial animals recovered from the Beltrán Onofre Banegas-Lami Hernández (BOL) site from the Chaco-Santiagueña archaeological region (province of Santiago del Estero), corresponding to a late agro-pottery stage. This is the first site excavation in the region undertaken within a taphonomic framework. Hence, it was possible to gather different taphonomic variables to assess the role of different taxa in the archaeological record.

In previous works, the subsistence system of the Chaco-Santiagueña region for the agro-pottery stage was characterized (del Papa 2012; del Papa et al. 2012). It is observed that ancient inhabitants used camelids (wild and domesticated) as a main resource throughout the analyzed period. In some cases, as a specialist strategy and, in some others, as a more generalist one together with other resources such as fish, *Rhea americana*, Dolichotinae rodents, Cervidae and Tayassuidae depending on the analyzed site. It is important to take into account that the use of some resources was occasional or seasonal such as *Tupinambis* sp., small and medium birds, *R. americana* eggs, Dasypodidae, Carnivora and possibly Caviinae.

9.1.1 Beltrán Onofre Banegas-Lami Hernández Site

The BOL archaeological site (Dept. Robles, Santiago del Estero), is located at 27° 49′ 08″ S and 64° 02′ 43″ W (Fig. 9.1) (del Papa and De Santis 2015). The settlement would have been more extensive than at present because plowing and land preparation destroyed part of the site (del Papa and De Santis 2015; Fig. 9.2). Archaeological excavations were conducted on the sector that showed no evidence of current anthropic modifications and where the settlement physiognomy could be partially seen (del Papa and De Santis 2015).

The site is located in the semi-arid subtropical and continental area of the country. The high average temperature (20 °C), the annual precipitation concentrated in the summer season (550 mm) and the great potential evapotranspiration capacity, determine local water deficiency for the study area (Ledesma 1979).

The ancient occupation is distinguished by the presence of mounds in the vicinity of a paleobed of the Dulce river. The archaeofaunal assemblage comes

Fig. 9.1 Location of the Beltrán Onofre Banegas-Lami Hernández site

from 10 grids with variable dimensions, 40 cm of deposits, and an excavated area of 28.75 m^2 (del Papa and De Santis 2015; Fig. 9.2). Sediments containing the archaeological remains were homogeneous, of a silt-sandy composition. Ceramic sherds were recovered during the field work, which after a preliminary study, were assigned to the late agro-pottery. There was a presence of mainly Sunchitúyoj and a few Averías ceramic fragments. Human bones (a few skull fragments) and lithic remains, mostly flakes (N = 23), followed by pebbles (N = 18) and projectile points (N = 4) were also recovered. Most of the remains recovered from the site (except archaeofaunal) were only preliminarily examined; an exhaustive analysis is yet to be undertaken. The site yields a radiocarbon date from charcoal samples of 420 ± 60 BP (LP-2054) (del Papa and De Santis 2015).

9.1.2   Cultural Context of the Study Site

The Late Agro-pottery period near the middle basin of the Dulce river is characterized by the presence of sedentary groups with a mixed economy (agriculture, livestock farming, hunting and gathering), settlement pattern of residential buildings on mounds (natural, artificial or mixed) and ceramic technology (Togo 2004). According to new available dates, the Late period would have developed the ceramic styles Sunchitúyoj (ca. 1200–1500 AD) and Averías (close to the arrival of European settlers in the sixteenth century) (Togo 2007). In many sites of the area these styles are to a greater or lesser extent associated, while there are also pure and contemporary occupation sites without any association (Gramajo de Martínez Moreno 1978; Reichlen 1940; Togo 2004). Lorandi (1978) highlights the similarities

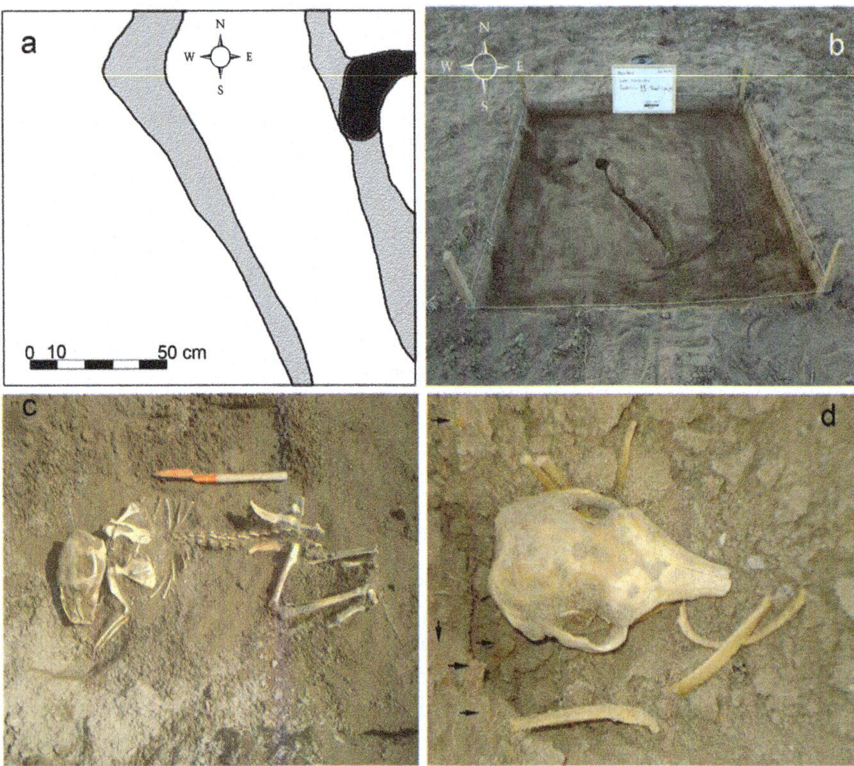

Fig. 9.2 Caves recorded during excavation and individuals found as articulated skeletons. (**a**) Grid number 8 diagram, burrow trajectory in *light grey* and deeper burrow in *dark grey*; (**b**) Photo of grid number 12, (**c**) *Lagostomus maximus*, (**d**) *Chaetophractus vellerosus*

regarding the settlement pattern and economic system of the groups sharing these styles, and mentions that their main differences would be in the intensity of some processes (e.g., more emphasis in textile practices and a population increase in Averías) and in decorative and stylistic aspects of pottery (Gramajo de Martínez Moreno 1978; Lorandi 1978; Togo 2004). It is noteworthy that for the Late Period there is a greater interaction with other regions of the country, which has been attributed to different processes (e.g., Bleiler 1948; del Papa 2012; Gramajo de Martínez Moreno 1978; Taboada et al. 2013; Togo 2004).

9.1.3 Fossorial Animals in the Assemblage

Here we consider fossorial fauna as those species which inhabit underground burrows during some time in their lives. Reptiles are ectothermal animals, therefore seasonal and daily thermal differences affect their behavior. In this sense, reptile activity in the studied region is mainly detected during the trophic period (from

September to April) when they are likely to be captured by humans and other predators (Richard 1999). By contrast, during the winter season (April–August) they stay in their refuges where they will hibernate (Richard 1999). Lizards of genus *Tupinambis* can build their own burrows or inhabit those abandoned by other animals (e.g., *Lagostomus maximus*; Fitzgerald 1992). Likewise, *Chelonoidis chilensis* tortoises also inhabit burrows abandoned by other animals (e.g., vizcachas, armadillos) (Richard 1999). More than 30 poisonous, constrictor and other ophidian species are listed in Santiago del Estero (Basualdo et al. 1985). Ophidians present very different patterns regarding hibernation habits (they can hibernate in trees, under stones or in burrows).

Most dasypodids are adapted to fossorial life and use their burrows to protect themselves from extreme temperature changes or as a defensive strategy against their predators (Redford and Eisenberg 1992). Among the species recorded at the site including *Cabassous chacoensis* (Chacoan naked-tailed armadillo), *Chaetophractus vellerosus* (screaming hairy armadillo) and *Tolypeutes matacus* (southern three-banded armadillo), only the latter does not dig its own burrow but reuses caves made by other armadillos or other burrowing mammals (Redford and Eisenberg 1992; Smith 2007).

Amongst the rodents, *Lagostomus maximus* includes colonial, gregarious and fossorial habits. They made extensive cave systems (burrows) called vizcacheras (Jackson et al. 1996) with multiple grouped entrances and where several individuals coexist (Redford and Eisenberg 1992).

The Dolichotinae rodents (*Dolichotis patagonum* and *Dolichotis salinicola*) often use the burrows to protect the young, where more than 22 couples often pool their young in a communal well (Redford and Eisenberg 1992). Some Caviinae, like *Microcavia australis* (small cavy) usually build burrow systems with multiple entries at shallow depth, where between 4 and 38 individuals can live (Tognelli et al. 2001). *Galea leucoblephara* (common cavy) constructs shallow galleries and can use the burrows of Ctenomyidae rodents. Ultimately, rodents of the *Ctenomys* genus are adapted to fossorial life and spend most of their lives under the soil surface (Redford and Eisenberg 1992). Their long and complex tunnel system contains storage chambers for both food and nest (Redford and Eisenberg 1992).

9.2 Methodology

During fieldwork contextual analyses were performed to identify burrows and to record skeletal articulation, for interpreting natural death (e.g. disease, old age, starvation, burrow collapse) and the spatial disturbance of fossorial animals. The high activity of excavating animals on the site may have affected the identification of articulated individuals. The differential distribution of chemical traces on skeletal elements (in this case manganese oxide patina) was analyzed to infer the death of articulated individuals. The absence of patina in joint areas of the elements would indicate that they were articulated when chemical deposition began.

The age profile of Dolichotinae rodents was estimated to infer whether it corresponds to the natural death of young individuals in their burrows. This analysis is complemented by contextual information and data obtained from the study of bone modifications, since the presence of young individuals may be related to human prey preference and probable capture technique (del Papa et al. 2010; Jones 2006). Age was calculated by the percentage of fused and unfused specimens for the proximal humeral and distal femoral epiphyses (del Papa et al. 2010; and references cited therein). Materials were quantified taking into account the following measures of taxonomic abundance and skeletal parts: NISP, NISP%, MNI, and MNE (following Mengoni Goñalons 1999). In order to contextualize the sample of burrowing animals with the whole archaeofaunal assemblage, bone surface modifications relating to natural processes of deposition and post-deposition, like rodents, root etching and weathering (e.g., Behrensmeyer 1978; Lyman 1994) were analyzed. Bone modifications generated by natural predators, either by mechanical action or by gastric acid corrosion (Andrews 1990; Binford 1981) were also analyzed. The gastric corrosion degree has been recorded in tooth elements, proximal femoral epiphysis and distal humeral epiphysis (Andrews 1990; Fernández-Jalvo and Andrews 1992; Gómez 2007). Moreover, changes in the bone surface were identified by human activity, especially taking into account cut marks (Blumenschine et al. 1996; Mengoni Goñalons 1999) and thermal alteration (Shipman et al. 1984; Stiner et al. 1995).

9.3 Results

9.3.1 General Characteristics of the Sample

Based on NISP%, the archaeofaunal assemblage is predominated by mammals (53.45%), followed by fish (29.43%), birds (9.33%), and reptiles (5.87%) (Table 9.1). Mammalian faunas are predominated by Dasypodidae, followed by *Lama* sp., *Lagostomus maximus* and Sigmodontinae, and Dolichotinae (*D. patagonum* and *D. salinicola*) and *Ctenomys* sp. (Table 9.1).

Among the naturally occurring bone modifications that affected the total archaeofaunal assemblage, precipitation of manganese oxide was the most prevalent (11.07% of specimens). Low proportions of specimens showed root (2.4%), rodent (0.48%) and carnivore marks (0.01%). A relatively quick burial of the materials is inferred by the higher proportion of mammal specimens larger than 5 kg with minimal (stages 0 and 1 = 78%) and middle stage (stages 2 and 3 = 20.4%) weathering. As much as 13.37% of the microvertebrate sample (frogs and micromammals) displays evidence of gastric corrosion; these are still being analyzed.

Table 9.1 Taxonomic abundance. NISP, MNI and NISP%

Taxa	NISP	NISP%	MNI	Weight (kg)[a]
Mollusca	176	1.69	47	–
Teleostei	3067	29.39	60	0.968[b]
Anura	23	0.22	2	–
Reptilia indet.	7	0.07	–	–
Chelonoidis chilensis	98	0.94	2	2.5
Ophidia	73	0.75	1	–
Tupinambis sp.	433	4.15	7	4.15
Iguania	7	0.07	1	–
Aves[c]	421	4.03	11	–
Rhea americana	71	0.68	4	26
Rhea americana (egg shells)	483	4.63	–	–
Mammalia indet.	3350	32.10	–	–
Marsupialia	2	0.02	1	–
Dasypodidae	112	1.07	–	–
Chaetophractus vellerosus	751	7.20	8	0.837
Cabassous chacoensis	14	0.13	1	3.5
Tolypeutes matacus	39	0.37	1	1.1
Carnivora	1	0.01	–	–
Canidae	3	0.03	1	–
Felidae	2	0.02	–	–
Oncifelis geoffroyi	2	0.02	1	3.59
Artiodactyla	275	2.64	–	–
Lama sp.	510	4.89	7	95
Cervidae	2	0.02	–	–
Mazama sp.	4	0.04	1	23.45
Rodentia	21	0.20	–	–
Lagostomus maximus	136	1.30	5	6.18
Dolichotinae	8	0.08	–	–
Dolichotis patagonum	58	0.56	3	10
Dolichotis salinicola	6	0.06	1	1.85
Caviinae	42	0.40	–	–
Galea leucoblephara	4	0.04	2	0.286
Microcavia australis	5	0.05	2	0.225
Ctenomys sp.	73	0.70	7	0.2
Sigmodontinae[d]	152	1.46	24	–
NISP Total	10,431	100	–	–
Indet.	5531			
NSP	15,962			

[a]Taxa average live weight

[b]For fish the average of the most conspicuous species of Dulce river was calculated

[c]Includes Tinamidae, Anseriformes, Anatidae, Falconiformes, Columbiformes, *Tyto alba* and Passeriformes

[d]Includes Akodontini, *Calomys callosus* and *Graomys* cf. *G. griseoflavus*

9.3.2 Context Information, Evidence of Buried Articulated Elements

During fieldwork, numerous animal caves made were noted in all the grids. With the exception of some (Fig. 9.2a, b), most of the caves could not be observed to their full extent since many of their sections were collapsed. During excavation three individuals of *C. vellerosus* (NISP = 345), one of *L. maximus* (NISP = 85) and one of *Ctenomys* sp. (NISP = 27) were found articulated (Fig. 9.2c, d; Table 9.2). Moreover, the analysis of differential representation of manganese oxide patina on elements allowed us to identify two individuals of *C. vellerosus* (NISP = 79) and remains of Dasypodidae (probably belonging to *C. vellerosus*; NISP = 20) has been buried while articulated (Fig. 9.3, Table 9.2).

9.3.3 Bone Modification, Evidence of Predator and Anthropic Accumulation

9.3.3.1 *Chelonoidis chilensis*

A total of 98 specimens of *C. chilensis* were recovered (MNI = 2 calculated on the recovery of two left ilia) (Table 9.1). More carapace plates (NISP = 83, MNE = 41) are observed in comparison to endoskeletal elements (NISP = 13, MNE = 11). High fragmentation of exoskeletal elements (75.29%) compared to the endoskeleton (30.77%) is also associated. Thermal alteration of carapace plates was observed, mainly on their dorsal sides (Fig. 9.4a, Table 9.2) as a possible product of cooking on embers or direct fire (del Papa and De Santis 2015). These results are consistent with signatures of anthropic consumption (del Papa and De Santis 2015; Sampson 2000; Thompson and Henshilwood 2014).

9.3.3.2 *Tupinambis* sp.

From the total of 433 specimens were identified as *Tupinambis* (MNI = 7 based on the jugal and maxilla). Most skeletal elements are observed, suggesting the deposition of complete individuals, and predominated by skull and vertebrae, followed by jaws and ribs elements, and to a lesser extent, girdle and limb elements (Table 9.3).

Evidence for anthropic accumulation of *Tupinambis* sp. includes seven elements with cut marks (1.61%), including the skull (two articular and three quadrates, Fig. 9.4b), a vertebra and a rib (del Papa 2015). Furthermore, 5.54% of *Tupinambis* sp. specimens show high degree thermal alteration (50% carbonized and 37.5% calcined) on elements from different regions of the skeleton (Table 9.2); a cooking pattern could not be observed.

9.3.3.3 Ophidia

Element representation is dominated by the vertebrae (MNE = 69) and a low proportion of ribs (MNE = 3) and skull (one maxilla). No evidence for natural death or the action of natural predators was found. Moreover eight vertebrae were

Table 9.2 Taxonomic abundance of fossorial fauna

Taxa	Quantity		Articulated		Gastric acid		Anthropic evidence					
	NISP	MNI	NISP	MNI	NISP	%[a]	Cut	%[a]	BT	%[a]	Burn	%[a]
Chelonoidis chilensis	98	2	–	–	–	–	–	–	–	–	31	31.63
Tupinambis sp.	433	7	–	–	–	–	7	1.61	–	–	24	5.54
Ophidia	73	–	–	–	–	–	–	–	–	–	8	10.95
Dasypodidae	112	–	20	–	–	–	–	–	–	–	4	4.49
Chaetophractus vellerosus	751	8	424	5	–	–	–	–	–	–	25	7.64
Tolypeutes matacus	39	1	–	–	–	–	–	–	–	–	13	35.13
Cabassous chacoensis	14	1	–	–	–	–	–	–	–	–	1	7.14
Ctenomys sp.	73	7	27	1	7	15.55	–	–	–	–	–	–
Lagostomus maximus	136	4	85	1	–	–	3	5.88	1	1.96	4	7.84
Dolichotinae	8	–	–	–	–	–	–	–	–	–	–	–
Dolichotis patagonum	58	3	–	–	–	–	4	6.89	1	1.72	10	17.24
Dolichotis salinicola	6	1	–	–	–	–	–	–	–	–	2	33.33
Caviinae	42		–	–	3	7.14	–	–	–	–	7	16.66
Galea leucoblephara	4	2	–	–	1	25	–	–	–	–	–	–
Microcavia australis	5	2	–	–	–	–	–	–	–	–	–	–
Total	1852		556		11		14		2		129	

Specimens with evidence of natural death in their burrows, evidence of the action of natural predator and specimens with anthropic signals

Cut specimens with cutmarks, *BT* bone tools, *Burn* burned specimens

[a]Specimens of individuals which died in their burrows are not taken into account (found in articulated state or from their interpretation by manganese oxide patinas)

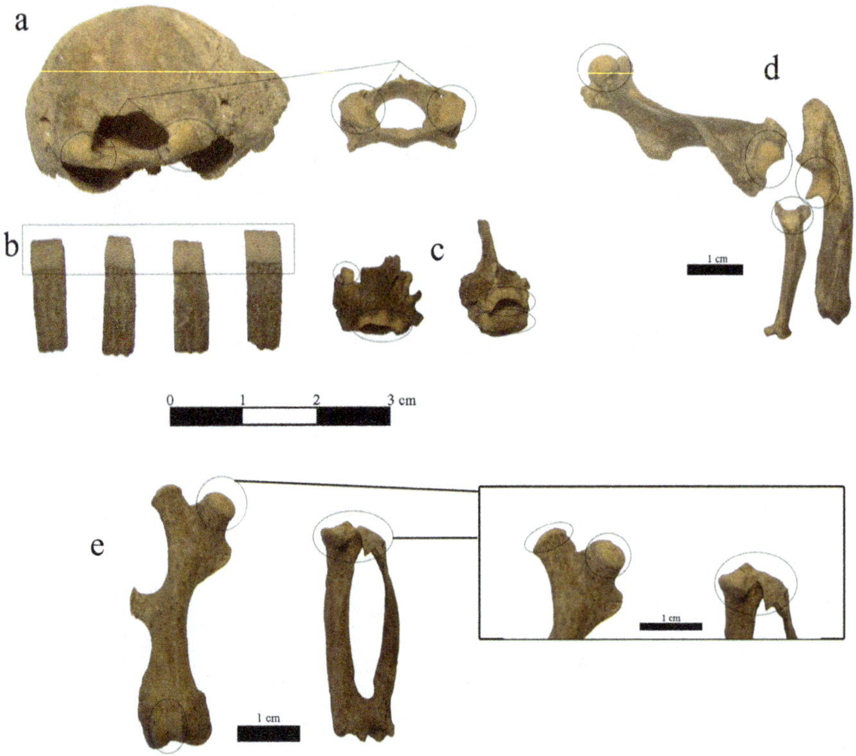

Fig. 9.3 *Chaetophractus vellerosus* elements with differential distribution of manganese oxide patinas. (**a**) Skull-atlas articulation; (**b**) mobile osteoderms; (**c**) thoracic vertebrae; (**d**) forelimb; (**e**) hindlimb

recovered with thermal alteration (10.95%), seven are carbonized and one is calcined (Table 9.2).

9.3.3.4 Dasypodidae

Not counting those individuals found in an articulated state, Dasypodidae specimens included elements of the endoskeleton (NISP = 89), mainly vertebrae (NISP = 30) and ribs (NISP = 13). Next in proportion are elements of girdles and limbs (NISP between 3 and 8 per element). Only 4.49% of the Dasypodidae specimens are thermally altered (NISP = 4, three carbonized and one calcined), including two vertebrae, one distal humeral epiphysis and one distal femur fragment. At the species level, bony armour plates predominate (*C. vellerosus*, NISP = 315; *C. chacoensis*, NISP = 14; *T. matacus*, NISP = 37) and to a lesser extent endoskeleton elements such as mandibles (NISP = 5), skull (NISP = 3), femur (NISP = 2) and humerus (NISP = 1) for *C. vellerosus*, and a mandible and radius for *T. matacus*. Regarding anthropic modification, a low percentage of thermally altered elements (Table 9.2) was observed, especially bony armour plates

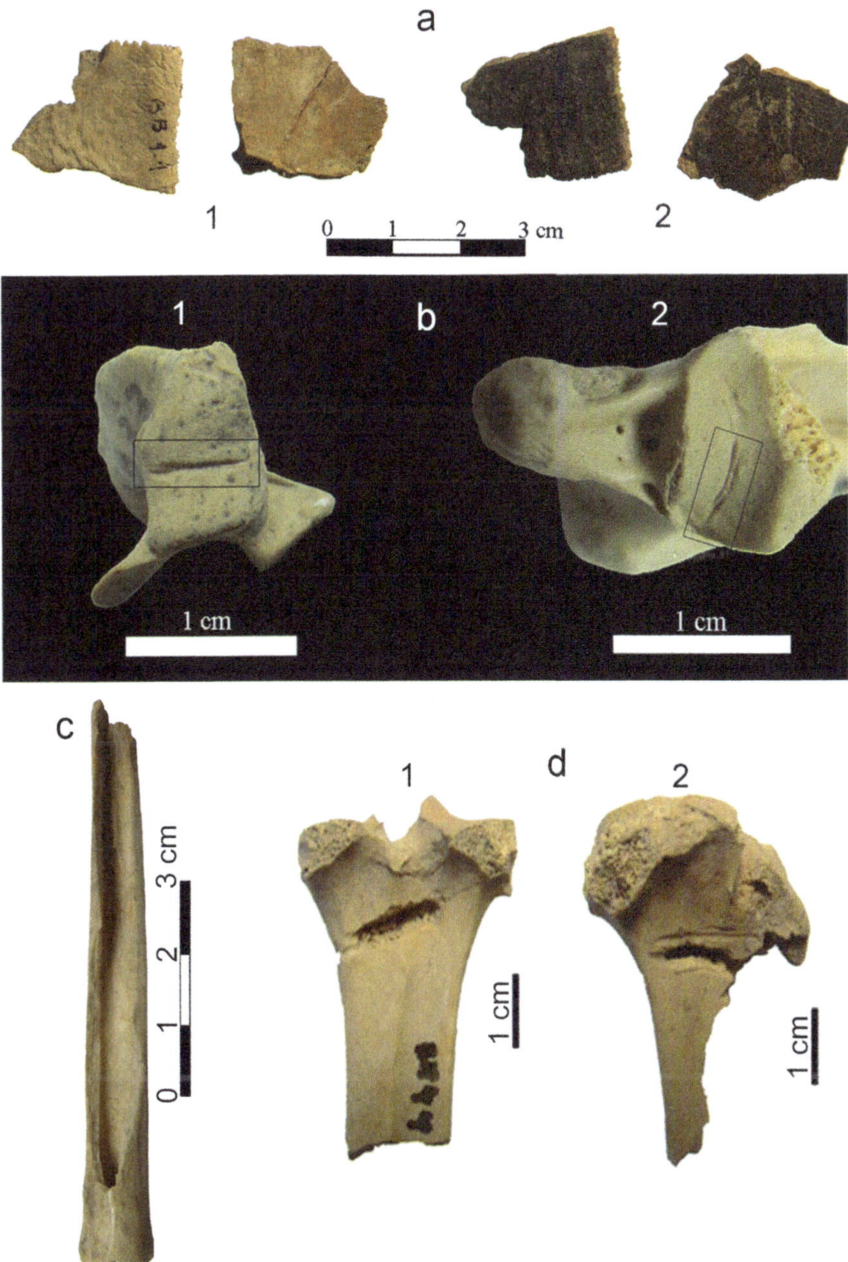

Fig. 9.4 Anthropic evidence. (**a**) *Chelonoidis chilensis* carapace plates, a1—ventral face without thermal alteration; a2—carbonized dorsal side; (**b**) Cut marks on specimens of *Tupinambis* sp., b1—Quadrate, ventral view, b2—Articular, occlusal and posterior view; (**c**) projectile point performed on *Lagostomus maximus* tibia; (**d**) proximal portion of tibia of *Dolichotis patagonum* with cut marks on the epiphysis, d1—posterior view, d2—lateral view

Table 9.3 Representation of skeletal elements (NISP and MNE) of *Tupinambis* sp. and rodents

	Tupinambis sp.		*Ctenomys* sp.		*L. maximus*		Dolichotinae		*D. patagonum*		*D. salinicola*		Caviinae	
Skeletal element	NISP	MNE	NISP	MNE	NISP	MNE	NISP	MNE	NISP	MNE	NISP	MNE	NISP	MNE
Cranium	110	7	8	4	6	2	4	1	3	1	–	–	10	2
Mandible	34	11	17	12	4	3	–	–	–	–	–	–	2	2
Incisors	–	–	2	–	1	–	–	–	–	–	–	–	–	–
Molars	–	–	3	–	–	–	1	–	1	–	–	–	8	–
Vertebrae	182	172	–	–	1	1	1	1	8	8	–	–	–	–
Ribs	26	26	–	–	–	–	–	–	–	–	–	–	–	–
Scapular girdle	7	4	–	–	3	3	–	–	1	1	–	–	–	–
Humerus	9	7	2	2	6	6	1	1	2	1	1	1	8	6
Radius	8	8	–	–	–	–	–	–	4	3	1	1	–	–
Ulna	1	1	1	1	1	1	–	–	2	2	2	2	–	–
Pelvic girdle	19	6	3	3	2	1	–	–	3	3	–	–	1	1
Femur	10	5	5	5	6	4	–	–	6	2	–	–	7	4
Patella	–	–	–	–	1	1	–	–	1	1	–	–	–	–
Tibia	8	5	5	5	6	4	1	1	7	4	–	–	6	5
Fibula	4	4	–	–	–	–	–	–	–	–	–	–	–	–
Basipodium bones	4	4	–	–	8	8	–	–	9	9	2	2	–	–
Metapodials	10	10	–	–	4	4	–	–	8	5	–	–	–	–
Phalanx	1	1	–	–	2	2	–	–	3	3	–	–	–	–
Total	433	271	46	32	51	40	8	4	58	43	6	6	42	20

of *C. vellerosus* (NISP = 24), *C. Chacoensis* (NISP = 1) and *T. matacus* (NISP = 13), 50% of which were burned exclusively on the dorsal side. A *C. vellerosus* femur was also observed with a thermally altered greater trochanter showing possible cooking. The greater preponderance of thermally altered Dasypodidae bony armour plates is suggestive of ember cooking based on experimental studies (Frontini and Vecchi 2014).

9.3.3.5 *Lagostomus maximus*

Not counting articulated individuals, *L. maximus* (MNI = 4 based upon four right humeri) include most portions of the skeleton, especially skull and limb specimens, and a virtual absence of the spinal column and ribs (Table 9.3). It is possible that these undiagnostic specimens are included as indeterminate small mammal remains. 5.88% of the specimens have cut marks (N = 3) on hind limb elements. One tibia was transformed into a projectile point (*punta semiacanalada ahuecada sin epífisis*, sensu Pérez Jimeno and Buc 2010), which presents an edge-bevel fracture at the distal end and retains a portion of the circumference of the tibial shaft at the proximal end. It is hollowed and does not retain the distal epiphysis (cut by perimeter sawing); the distal end of the device is fragmented (Fig. 9.4c, Table 9.2). 7.84% of the specimens display thermal alteration; predominated by carbonized distal elements of the limbs (an astragalus, a phalanx and a basipodium bone), and a partially burned supraoccipital fragment. The thermal alteration of distal limb elements is consistent with the pattern observed by Medina et al. (2012) for Caviomorpha rodents cooked in embers or direct fire.

9.3.3.6 Dolichotinae

At the subfamily level, the fragmentary remains were assigned mainly to crania (NISP = 5), cervical vertebra, humerus and tibia (NISP = 1, respectively) specimens. To *D. patagonum* were identified from both elements of the axial and appendicular skeleton; vertebrae and limb specimens predominate (Table 9.3); a MNI = 3 is based on the proximal epiphysis of the radius (two merged right and an unfused left). With regard to the age of the individuals present in the collection, only three specimens were taking into account (two distal femoral epiphysis and a proximal humeral epiphysis) which are juveniles since they were unfused. 6.89% of *D. patagonum* specimens have cut marks: an epiphysis of a femur, proximal epiphysis of a radius, a calcaneum and a metapodial. In 1.72% of the sample, early-stage manufacture of an artifact appears on the proximal half of a tibia which has transverse and deep incisions in both its posterior, medial and lateral side, with the goal of controlling fracture of the element (Fig. 9.4d, Table 9.2). 17.24% of the remains are thermally altered to a high degree (60% carbonized and 20% calcined). A proximal portion of a femur and a distal portion of a tibia have a low degree of alteration and may represent cooking. Few elements of the limbs of an individual have been identified as *D. salinicola* (Table 9.3). A distal humeral epiphysis and a half proximal ulna are carbonized.

9.3.3.7 Caviinae

At a subfamily level fragmentary remains of skull and limb's elements (Table 9.3) were recognized. At a species level cranial elements (three mandibles and a maxilla of *G. leucoblephara*, and three mandibles and two maxilla of *M. australis*) were assigned.

7.14% of Caviinae remains and 25% of *G. leucoblephara* displayed gastric corrosion (Table 9.2) with light grade. Gastric corrosion was recorded in a femur, a humerus, maxillary molars of Caviinae and maxillary molars of *G. leucoblephara*. These specimens suggest the action of natural predators. A small percentage of Caviinae remains display thermal alteration (16.66%, Table 9.2) on different elements of the skeleton (skull, mandibles, femur, humerus and tibia). Only a tibia displays the pattern of ember cooking observed by Medina et al. (2012) with thermal alteration in its distal portion; however, it is insufficient to affirm this activity.

9.3.3.8 *Ctenomys* sp.

Besides the remains recorded as natural death in their burrows (see Sect. 9.3.2), 15.55% of *Ctenomys* sp. specimens display gastric corrosion (Table 9.2). The light grades of corrosion stand out, observed in a femur and molars of two mandibles and a maxilla. Only one distal epiphysis of a humerus displays moderate grade. These specimens suggest the action of natural predators.

9.4 Discussion and Conclusions

One of the primary objectives of zooarchaeology since its inception has been to identify the accumulating agent of bone recovered at archaeological sites on the basis of taphonomic signatures (Binford 1981). In this regard, the excavation of a site from a taphonomic framework enables us to distinguish the different agents of accumulation, and thus allows us greater accuracy in the archaeological interpretation and the estimation of the dietary roles for different resources. The activity of burrowing animals (i.e. faunalturbation) complicates the task of identifying accumulator agents from archaeological remains (Shaffer 1992; Stahl 1996). This faunalturbation not only can affect horizontal and vertical distribution of remains (e.g. Bocek 1986; Erlandson 1984), but also can modify the previous faunalturbation (Erlandson 1984), as it has been observed in BOL site throughout the interpretation of natural death of individuals in their burrows by differential pattern of chemical deposition.

This work contains preliminary results obtained from a site excavated within a taphonomic framework in the Chaco-Santiagueña region. It enabled us to differentiate between the deposition of fossorial animals that entered into the record due to the natural death of individuals in their burrows, the contribution of natural predators and anthropic incorporation. In this regard, it is suggested that the mere analysis of bone modifications in order to interpret agents that contributed to the archaeological record is insufficient (e.g. Lyman 1994; Morlan 1994; Stahl 1996;

Weissbrod and Zaidner 2014). An example of this can be observed in previous studies for the region (del Papa 2012), where in the absence of evidence for non-human predation alongside the presence of some specimens of fossorial fauna with direct or indirect evidence of human activity it was suggested that all represented individuals could be the result of human activity (e.g., *L. maximus*, Dasypodidae; del Papa 2012; Togo 2004).

Our analysis, integrating site context evidence, skeletal element representation and bone modifications, allows us to verify that the preserved remains of *C. chilensis*, *Tupinambis* sp., and Dolichotinae rodents are mostly the result of human activity; cut marks (*Tupinambis* sp. and *D. patagonum*), initial stages in the production of artifacts (*D. patagonum*) and thermal alteration (*C. chilensis*, *Tupinambis* sp., and Dolichotinae) are identified. It is noteworthy that cooking of *C. chilensis*, *T. matacus* and *D. patagonum* was inferred from the pattern of thermal alteration. Note that in the case of *T. matacus* this cooking pattern has been observed in other sites of the study region (del Papa 2012). Although the high proportion of *D. patagonum* juveniles (taking into account the specimens that allow us to estimate age) could indicate the accumulation of several individuals by natural death in their burrows, the absence of articulated individuals, and the presence of cut marks, thermal alteration and initial stages in the production of artifacts suggest the anthropic accumulation may be through hunting from a communal well (mating pairs usually deposit their juveniles in such wells; Redford and Eisenberg 1992). However, since the number of specimens that allow us to estimate age (proximal humeral and distal femoral epiphysis) are scarce, and in the absence of a detailed report on their age development, this inference is inconclusive pending more study (del Papa et al. 2010).

The representation of skeletal parts of *C. chilensis* and *Tupinambis* sp. is consistent with the entrance of skeletally complete individuals into the site. For the case of both *T. matacus* and Dolichotinae rodents, the absence of some elements could be related to recovery biases, or to difficulty in the systematic identification of fragmentary remains due to butchering or post-depositional processes, resulting in their inclusion into more comprehensive taxonomic categories (e.g., Mammalia indet., Dasypodidae and Rodentia).

Moreover, for both *L. maximus* and *C. vellerosus* there was evidence of natural death of individuals in their burrows for some, and the anthropic incorporation of others (cut marks and manufacture of artifacts for *L. maximus* and thermal alteration inferring ember cooking or direct heat for both species).

Regarding Ophidia, *C. chacoensis* and Caviinae it is difficult as yet to infer whether their remains were incorporated by natural or cultural causes. This is due to the low representativeness in the record and the absence of direct evidence of accumulation (they present a low proportion of thermally altered elements—indirect evidence—without representing a cooking pattern), and in the case of Caviinae, they also show evidence of possibly Strigiformes action by the presence of remains with low-grade gastric corrosion (Andrews 1990; Fernández-Jalvo and Andrews 1992; Gómez 2007).

Finally, *Ctenomys* sp. remains have been introduced by both the natural death in their burrows and the action of a natural predator possibly Strigiformes (they have the same characteristic as Caviinae remains). It is important to point out that the taphonomic analysis of microvertebrates remains (Anura, Iguania, Aves, Sigmodontinae and Marsupialia) is still in progress and will allow us to be more precise regarding the role that various taxa had in human diet or natural processes that contributed to the record.

The results obtained in this work allowed us to reach to a more precise inference regarding resource quantification and their importance. A comparison of NISP% was made between the taxa represented in the record before and after excluding the animals incorporated into the site by natural death (Fig. 9.5). A change in resources proportion has been observed. In this sense, it greatly reduces the prevalence of *C. vellerosus* in the assemblage. This reduction of *C. vellerosus* generates an increase in other resources, mainly fish. Even if there are a wide variety of resources in the site, results show a lesser diversity in their use than prior to the analysis herein performed (greater proportion in the use of few resources such as fish and camelids). Now, if resources weight is taken into account, camelids constitute the most important economic resource in the sample (Table 9.1) and the remaining resources are considered complementary and/or occasional. This agrees with the observations in the study region for the late agro-pottery sites (del Papa 2012). For such period it is considered that agriculture is more intensive that previous times

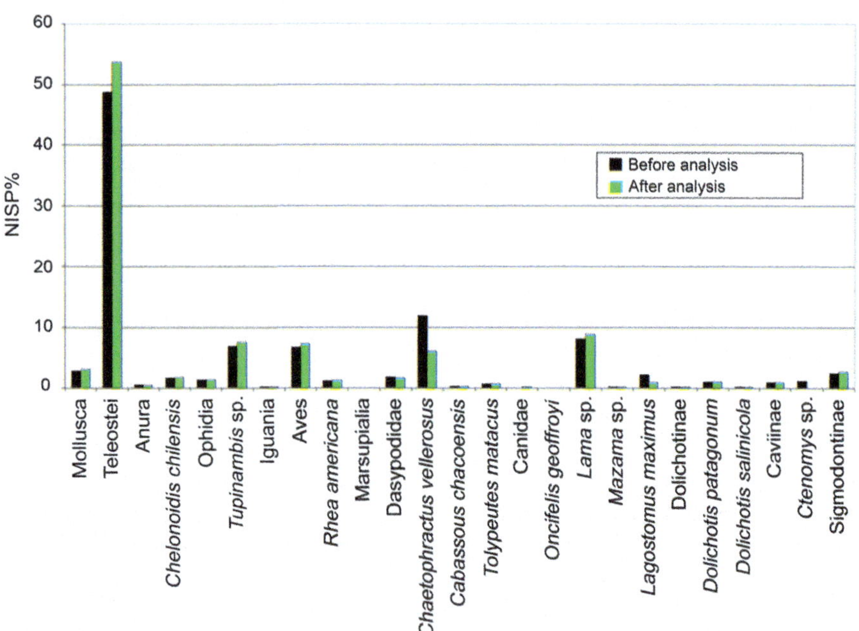

Fig. 9.5 Comparison of taxonomic abundance (NISP%) of the taxa represented in the record before and after excluding the animals incorporated into the site by natural death

and the surrounding settlements area depleted of wild resources. Hunters are forced to go across greater distances to look for resources, for that reason, preferred bigger preys, with greater energetic return rate or a greater use of domesticated animals (del Papa 2012; James 1990; Neme and Gil 2008; Szuter and Bayham 1989).

Acknowledgements To Valeria Accinelli for the English translation. To Professor Juan Carlos Cejas for facilitating fieldwork and his continued support for archaeological work in the area, to doña Elda and Mr. Lami Hernández for allowing us to work on their properties and providing their facilities, to the former Mayor of Beltrán, Miguel Alvarez for his support. To revisers Peter Stahl y Analía Andrade whose comments helped to improve the paper. What has been proposed in this paper is responsibility of their authors.

This work is part of the Postdoctoral fellowship from CONICET for the first author and is financed by The Programa del Proyecto de Incentivos for Teachers-Researchers, Facultad de Ciencias Naturales y Museo, UNLP, Code Project: 11/N601. Director: Dr. Luciano De Santis.

References

Acosta A, Pafundi L (2005) Zooarqueología y tafonomía de *Cavia aperea* en el humedal del Paraná inferior. Intersec Antropol 6:59–74

Albino AM (1999) Serpientes del sitio arqueológico Cueva Tixi (Pleistoceno tardío-Holoceno), Provincia de Buenos Aires, Argentina. Ameghiniana 36(3):269–273

Albino AM (2001) Reptiles. In: Mazzanti DL, Quintana C (eds) Cueva Tixi: Cazadores y recolectores de las Sierras de Tandilia Oriental. Tomo I. Geología, Paleontología y Zooarqueología. Laboratorio de Arqueología, Universidad Nacional de Mar del Plata, Mar del Plata, pp 65–74. Publicación Especial 1

Albino AM, Franco N (2011) Lagartijas (Iguania: Familia Liolaemidae) procedentes del sitio arqueológico Bi Aike Cueva 3 (Provincia de Santa Cruz, Argentina). An Inst Patagonia 39:127–131

Albino AM, Kligmann DM (2007) An accumulation of bone remains of two *Liolaemus* species in an Holocene archaeological site of the Argentinian Puna. Amphibia-Reptilia 28:154–158

Andrews PJ (1990) Owls, caves and fossils. University of Chicago Press, Chicago

Basualdo MA, Togo J, Urtubey N (1985) Aprovechamiento socioeconómico de la fauna autóctona de Santiago del Estero (inventario y uso popular más frecuente). Indoamérica 1. Publicación del Laboratorio de Antropología, Facultad de Humanidades, Universidad Nacional de Santiago del Estero, Santiago del Estero

Behrensmeyer AF (1978) Taphonomic and ecologic information from bone weathering. Paleobiology 4:150–162

Binford LR (1981) Bones: ancient men and modern myths. Academic Press, New York

Bleiler EF (1948) The east. In: Bennet WC, Bleiler EF, Sommer FH (eds) Northwest Argentine archaeology. Yale University Press, New Haven, pp 120–139

Blumenschine RJ, Marean CW, Capaldo SD (1996) Blind test of inter-analyst correspondence and accuracy in the identification of cut marks, percussion marks, and carnivore tooth marks on bone surfaces. J Archaeol Sci 23:493–507

Bocek B (1986) Rodent ecology and burrowing behavior: predicted effects on archaeological site formation. Am Antiq 51(3):589–603

Bond M, Caviglia SE, Borrero L (1981) Paleoetnozoología del Alero de los Sauces (Neuquén, Argentina); con especial referencia a la problemática presentada por los roedores en sitios patagónicos. Tr Prehistoria 1:95–111

Cione AL, Lorandi AM, Tonni EP (1979) Patrón de subsistencia y adaptación ecológica en la aldea prehispánica El Veinte, Santiago del Estero. Rel Soc Arg Ant 13:102–116 del Papa LM (2012) Una aproximación al estudio de los sistemas de subsistencias a través del análisis arqueofaunístico en un sector de la cuenca del Río Dulce y cercanías a la Sierra de Guasayán. PhD thesis, Facultad de Ciencias Naturales y Museo, UNLP, La Plata

del Papa LM (2015) Consumo de reptiles durante el período agroalfarero de la región Chaco-Santiagueña, Argentina. Archaeofauna 24:7–26

del Papa LM, De Santis LJM (2015) No se les escapó la tortuga. Uso antrópico de *Chelonoidis Chilensis* en un sitio de la región Chaco-Santiagueña (provincia de Santiago del Estero). Arqueología 21(1):115–135

del Papa LM, De Santis LJM, Togo J (2010) Consumo de roedores en el sitio Villa La Punta, agro-alfarero temprano de la región Chaco-Santiagueña. Intersec Antropol 11:29–40

del Papa LM, De Santis LJM, Togo J (2012) Zooarqueología santiagueña. Despertando de la siesta. In: Acosta A, Loponte D, Mucciolo L (comp) Temas de Arqueología. Estudios Tafonómicos y Zooarqueológicos (II). Asociación Amigos del Instituto Nacional de Antropología, Buenos Aires, pp 1–24

Erlandson JM (1984) A case study in faunalturbation: delineating the effects of the burrowing pocket gopher on the distribution of archaeological materials. Am Antiq 49(4):785–790

Fernández-Jalvo Y, Andrews P (1992) Small mammal taphonomy of Gran Dolina, Atapuerca (Burgos), Spain. J Archaeol Sci 19:407–428

Fitzgerald LA (1992) La historia natural de *Tupinambis*. Rev Un Nac Asunción 3:71–72

Fowler KD, Greenfield HJ, van Schalkwyk LO (2004) The effects of burrowing activity on archaeological sites: Ndondonwane, South Africa. Geoarchaeology 19:441–470

Frontini R, Deschamps C (2007) La actividad de *Chaetophractus villosus* en sitios arqueológicos. El Guanaco como caso de estudio. In: Bayón C, Pupio A, González MI et al (eds) Arqueología en Las Pampas. Sociedad Argentina de Antropología, Buenos Aires, pp 439–451

Frontini R, Escosteguy PD (2011) *Chaetophractus villosus*: a disturbing agent for archaeological contexts. Int J Osteoarchaeol 22:603–615

Frontini R, Vecchi R (2014) Thermal alteration of small mammal from El Guanaco 2 site (Argentina): an experimental approach on armadillos bone remains (Cingulata, Dasypodidae). J Archaeol Sci 44:22–29

Gómez G (2007) Predators categorization based on taphonomic analysis of micromammals bones: a comparison to proposed models. In: Gutiérrez MA, Miotti L, Barrientos G et al (eds) Taphonomy and zooarchaeology in Argentina, BAR international series 1601. Archaeopress, Oxford, pp 89–103

Gramajo de Martínez Moreno A (1978) Evolución cultural en el territorio santiagueño a través de la arqueología, Serie Monográfica No 5. Publicación del Museo Arqueológico "Emilio Y Duncan Wagner", Santiago del Estero

Jackson JE, Branch LC, Villarreal D (1996) Lagostomus maximus. Mamm Species 543:1–6

James S (1990) Monitoring archaeofaunal changes during the transition to agriculture in the American southwest. Kiva 56:25–43

Jones EL (2006) Prey choice, mass collecting and the wild European rabbit (*Oryctolagus cuniculus*). J Anthropol Archaeol 25:275–289

Ledesma NR (1979) La verdad sobre el clima de Santiago del Estero, Cuaderno de Cultura, vol 10 (17). Municipalidad de Santiago del Estero, Santiago del Estero

Lorandi AM (1978) El desarrollo cultural prehispánico en Santiago del Estero, Argentina. J Soc Am 65(1):63–85

Lyman RL (1994) Vertebrate taphonomy. Cambridge University Press, Cambridge

Medina M, Rivero D, Teta P (2011) Consumo antrópico de pequeños mamíferos en el Holoceno de Argentina Central: Perspectivas desde el abrigo rocoso Quebrada del Real 1 (Pampa de Achala, Córdoba). Lat Am Antiq 22:618–631

Medina M, Teta P, Rivero D (2012) Burning damage and small-mammal human consumption in Quebrada del Real 1 (Cordoba, Argentina): an experimental approach. J Archaeol Sci 39:737–743

Mello Araujo A, Marcelino JC (2003) The role of armadillos in the movement of archaeological materials: an experimental approach. Geoarchaeology 18:433–460

Mengoni Goñalons GL (1999) Cazadores de guanacos de la estepa patagónica. Colección tesis doctorales, Sociedad Argentina Antropología, Buenos Aires

Morlan RE (1994) Rodent bones in archaeological sites. Can J Archaeol 18:135–142

Neme GA, Gil AF (2008) Faunal exploitation and agricultural transitions in the South American agricultural limit. Int J Osteoarchaeol 18(3):293–306

Pardiñas UFJ, Teta P, Formoso AE, Barberena R (2011) Roedores del extremo austral: tafonomía, diversidad y evolución ambiental durante el Holoceno tardío. In: Borrero LA, Borrazo K (eds) Bosques, montañas y cazadores. Investigaciones Arqueológicas en Patagonia Meridional. Editorial Dunken, Buenos Aires, pp 61–84

Pérez Jimeno L, Buc N (2010) Tecnología ósea en la cuenca del Paraná. Integrando los conjuntos arqueológicos del tramo medio e inferior. In: Berón M, Luna L, Bonomo M et al (eds) Mamül Mapu: pasado y presente desde la arqueología pampeana. Libros del Espinillo, Ayacucho, pp 216–228

Politis G, Madrid P (1988) Un hueso duro de roer: Análisis preliminar de la tafonomía del sitio Laguna Tres Reyes (Partido de Adolfo González Chaves. Provincia de Buenos Aires). In: Ratto N, Haber A (eds) De procesos, contextos y otros huesos. Universidad de Buenos Aires, Buenos Aires, pp 29–44

Quintana CA, Mazzanti DL (2011) Las vizcachas pampeanas (*Lagostomus maximus*, Rodentia) en la subsistencia indígena del Holoceno tardío de las sierras de Tandilia oriental (Argentina). Lat Am Antiq 22:253–270

Quintana CA, Valverde F, Mazzanti DL (2002) Roedores y lagartos como emergentes de la diversificación de la subsistencia durante el Holoceno Tardío en sierras de la región Pampeana Argentina. Lat Am Antiq 13(4):455–473

Quintana CA, Mazzanti DL, Valverde F (2004) El lagarto overo como recurso faunístico durante el Holoceno de las sierras de Tandilia Oriental, provincia de Buenos Aires. In: Gradín C, Oliva F (eds) La Región Pampeana, su pasado arqueológico. Editorial Laborde, Buenos Aires, pp 347–353

Redford KJ, Eisenberg JF (1992) Mammals of the Neotropics. The southern Cone, vol 2. University of Chicago Press, Chicago

Reichlen H (1940) Reserches Archeologiques dans la province de Santiago del Estero (Rep. Argentine). J Soc Am 32:133–225

Richard E (1999) Tortugas de las regiones áridas de Argentina: Contribución al conocimiento de las tortugas de las regiones áridas de Argentina (Chelidae y Testudinidae) con especial referencia a los aspectos ecoetológicos, comerciales y antropológicos de las especies del complejo chilensis (*Chelonoidis chilensis* y *C. donosobarrosi*) en la provincia de Mendoza. L.O.L.A. Literature of Latin America Press, Buenos Aires

Salemme M, Escosteguy P, Frontini R (2012) La fauna de porte menor en sitios arqueológicos de la región Pampeana, Argentina. Agentes disturbadores vs recurso económico. Archaeofauna 21:163–185

Sampson CG (2000) Taphonomy of tortoises deposited by birds and Bushmen. J Archaeol Sci 27:779–788

Santiago FC (2004) Los roedores en el "menú" de los habitantes de Cerro Aguará (provincia de Santa Fe): su análisis arqueofaunístico. Intersec Antropol 5:3–18

Santini M (2012) Aprovechamiento de *Myocastor coypus* (Rodentia, Caviomorpha) en sitios del Chaco Húmedo argentino durante el Holoceno tardío. Intersec Antropol 12:195–205

Shaffer BS (1992) Archaeology interpretation of gopher remains from Southwestern archaeological assemblages. Am Antiq 57:683–691

Shipman P, Foster GF, Schoeninger M (1984) Burnt bones and teeth: an experimental study of colour, morphology, crystal structure and shrinkage. J Archaeol Sci 11:307–325

Smith P (2007) Southern three-banded armadillo. *Tolypeutes matacus* (Desmarest, 1804). In: Fauna Paraguay, handbook of the mammals of Paraguay, vol 7, pp 1–12. http://faunaparaguay.com/tolypeutesmatacus.html. Accessed 26 Nov 2014

Soibelzon E, Medina M, Abba AM (2013) Late Holocene armadillos (Mammalia, Dasypodidae) of the Sierras of Córdoba, Argentina: zooarchaeology, diagnostic characters and their paleozoological relevance. Quat Int 299:72–79

Stahl PW (1996) The recovery and interpretation of microvertebrate bone assemblages from archaeological contexts. J Archaeol Method Theory 3:31–75

Stiner MC, Kuhn SL, Weiner S, Bar-Yosef O (1995) Differential burning, recrystalization, and fragmentation of archaeological bone. J Archaeol Sci 22:223–237

Szuter C, Bayham F (1989) Sedentism and prehistoric animal procurement among desert horticulturalist of the North American Southwest. In: Kent S (ed) Farmers as hunters. Cambridge University Press, Cambridge, pp 80–95

Taboada C, Angiorama CI, Leiton DM, López Campeny S (2013) En la llanura y en los valles... Relaciones entre las poblaciones de las tierras bajas santiagueñas y el Estado Inca. Intersec Antropol 14:137–156

Thompson JC, Henshilwood CS (2014) Tortoise taphonomy and tortoise butchery patterns at Blombos Cave, South Africa. J Archaeol Sci 41:214–229

Tognelli MF, Campos CM, Ojeda RA (2001) *Microcavia australis*. Mamm Species 648:1–4

Togo J (2004) Arqueología Santiagueña: Estado actual del Conocimiento y Evaluación de un Sector de la Cuenca del Río Dulce. PhD thesis, Facultad de Ciencias Naturales y Museo, UNLP, La Plata

Togo J (2007) Los fechados radiocarbónicos de Santiago del Estero. In: Actas de resúmenes ampliados del XVI Congreso de Arqueología Argentina, vol 3, San Salvador de Jujuy, pp 227–232

Weissbrod L, Zaidner Y (2014) Taphonomy and paleoecological implications of fossorial microvertebrates at the Middle Paleolithic open-air site of Nesher Ramla, Israel. Quat Int 331:115–127

Wood WR, Johnson DL (1978) A survey of disturbance processes in archaeological site formation. In: Schiffer MB (ed) Advances in archaeological method and theory, vol 1. Academic Press, New York, pp 315–381

Archaeological Collagen Fingerprinting in the Neotropics; Protein Survival in 6000 Year Old Dwarf Deer Remains from Pedro González Island, Pearl Islands, Panama

10

Michael Buckley, Richard G. Cooke, María Fernanda Martínez, Fernando Bustamante, Máximo Jiménez, Alexandra Lara, and Juan Guillermo Martín

10.1 Introduction

Within the Neotropics ecozone, the tropical ecoregions of Central America host a unique collection of biodiversity that has been impacted greatly by human settlement. Due to major recent losses of diversity, studies that solely rely on the use of extant taxa provide a limited view of species change through human activity. Archaeological faunal remains can yield insights into this process of human-animal interactions in prehistory, but in many humid tropical regions these biological remains degrade relatively quickly rendering those that do survive more significant. This study looks at a Late Preceramic archaeological site occupied between 6190 and 5550 calibrated years before present (cal BP) (Table 10.1). This site is located on Pedro González Island, 8 km North of San José Island (Cooke and Jiménez 2009), Pearl Islands, Panama (Fig. 10.1), where the average temperature is >30 °C throughout the year. Osseous remains were found during fieldwork undertaken between 2008 and 2010 at the pre-Columbian site of Playa Don Bernardo (hereafter

M. Buckley (✉)
Faculty of Life Sciences, Manchester Institute of Biotechnology, University of Manchester, Manchester M1 7DN, UK
e-mail: m.buckley@manchester.ac.uk

R.G. Cooke • M.F. Martínez • M. Jiménez • A. Lara
Smithsonian Tropical Research Institute, P.O. Box 0843-03092, Balboa, Ancón, Republic of Panama

F. Bustamante
Universidad de Antioquia, Calle 67 Número 53-108, Medellín, Colombia

J.G. Martín
Universidad del Norte, Km 5 via Puerto Colombia, Barranquilla, Colombia

© Springer International Publishing AG 2017
M. Mondini et al. (eds.), *Zooarchaeology in the Neotropics*,
DOI 10.1007/978-3-319-57328-1_10

157

Table 10.1 Radiocarbon dates from the Preceramic shell bearing midden at Playa Don Bernardo, Pedro González Island, Pearl Island archipelago, Panama

Cut	Level (10 cm)	Stratum	Macrostratum	Material	Laboratory number	Radiocarbon age BP	$^{13}\delta$	Conventional age BP	Conventional age BC	Calibrated age (2σ)	Calibrated age (1σ)	Intercept
L-19	5		1	Marine bivalve (*Argopecten circularis*)	β-256752	4860 ± 50	−0.6	5260 ± 50	3310	5720–5560[a]	5650–5580	5600
L-19	8		1	Carbonized palm fruit	β-256751	4900 ± 40	−26	4880 ± 40	2930	5660–5880	5640–5590	5600
L-19	12		1	Marine bivalve (*Argopecten circularis*)	β-243898	4980 ± 40	0	5390 ± 40	3440	5870–5650[a]	5870–5650	5740
L-20	13		2	Charred plant matter	β-261219	5240 ± 50	−29	5170 ± 50	3220	6000–5890 and 5810–5760	5980–5970 and 5940–5900	5920
L-20	19		3	Charred plant matter	β-261218	5140 ± 40	-26	5120 ± 40	3170	5840–5750 and 5840–5750	5920–5890 and 5800–5770	5900
L-20	26		3	Charred plant matter	β-261217	5150 ± 40	−27	5120 ± 40	3170	5930–5850 and 5840–5750	5920–5890 and 5800–5770	5900
B'17	39	7	3	Charred dolphin (*Delphinus*) bone	β-304632	5350 ± 40	−14	5540 ± 40	3590	5990–5870[a]	5950–5890	5910
B'17	41	7	3	Charred plant matter	β-278902	5330 ± 40	−28	5280 ± 40	3330	6190–5930	6180–6150 and 6120–5990	6000

Dates and calibrations (INTEL-04) by Beta Analytic Inc.
[a]Marine calibration used

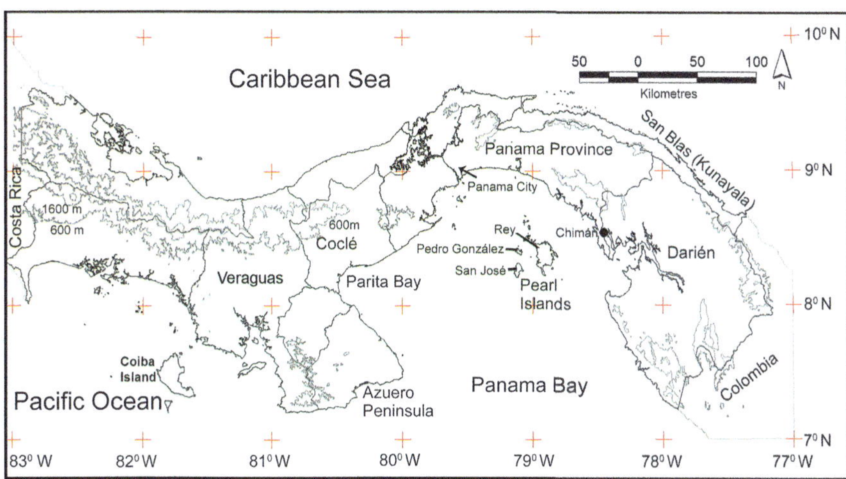

Fig. 10.1 Map of the Pearl Islands (*top*); Location of the Playa Don Bernardo (PDB) on Pedro González Island (*bottom*, *black arrow*; Photo by S. Redwood)

referred to as 'PDB'), at which the remains of a very small cervid were the most abundant in the terrestrial vertebrate sample (Martínez-Polanco et al. 2015).

In the PDB midden, 2502bone and antler specimens were assigned to Cervidae with a minimum number of individuals (MNI) of 22 (sum of MNI for each of three strata identified in each of three test pits) (Martínez-Polanco et al. 2015). The very high specimen/individual ratio across the site (114:1) is likely to be due, firstly, to the small size of the excavated test cuts, none of which uncovered complete cultural deposits, and, secondly, to the extreme fragmentation of the bones, which were heavily transformed by the actions of the pre-Columbian inhabitants while preparing the carcasses for food and tools (Fig. 10.2). A femur (TL: 132 mm) and a tibia

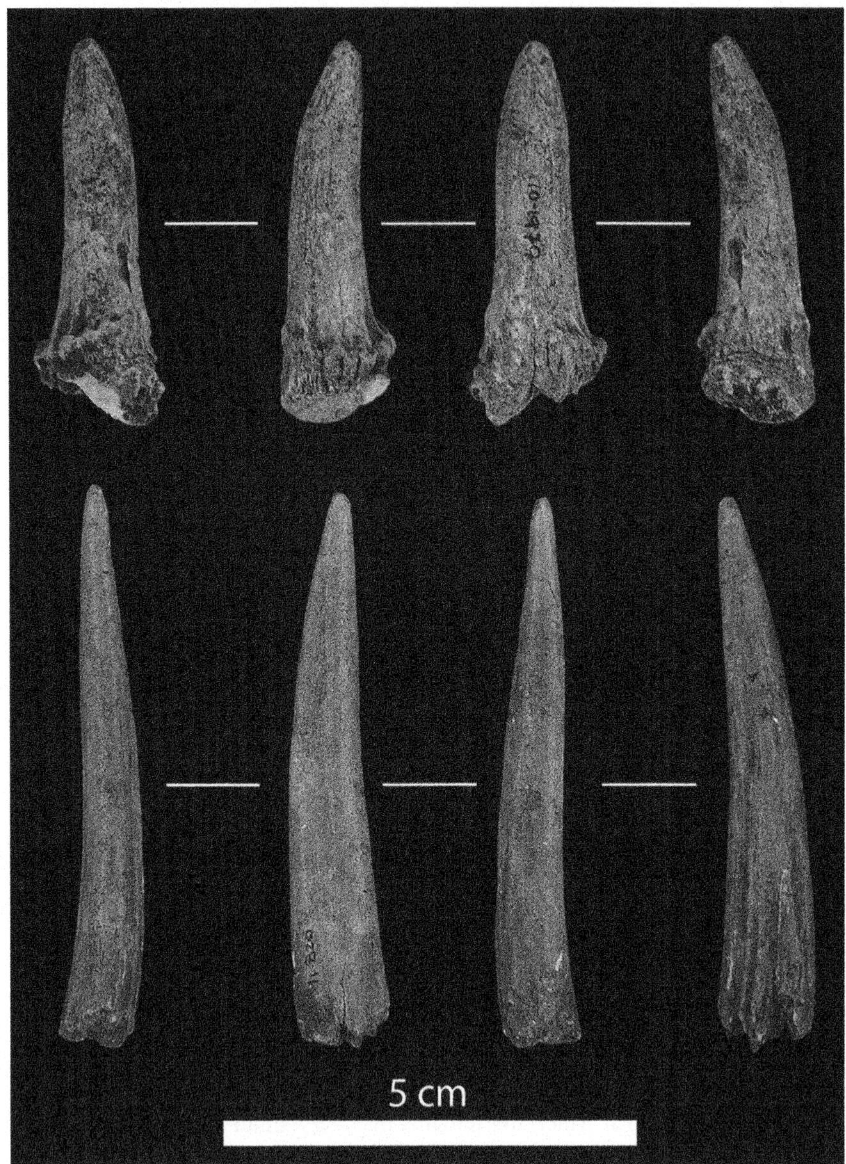

Fig. 10.2 Two deer antlers modified for use as tools, Playa Don Bernardo, Isla Pedro Gonzalez, Pearl Islands. *Top row*: Corte 1, Macrostratum IB (STRI cat.: 10–1979); *bottom row*: Corte 1, Macrostratum II (STRI cat.: 11–820)

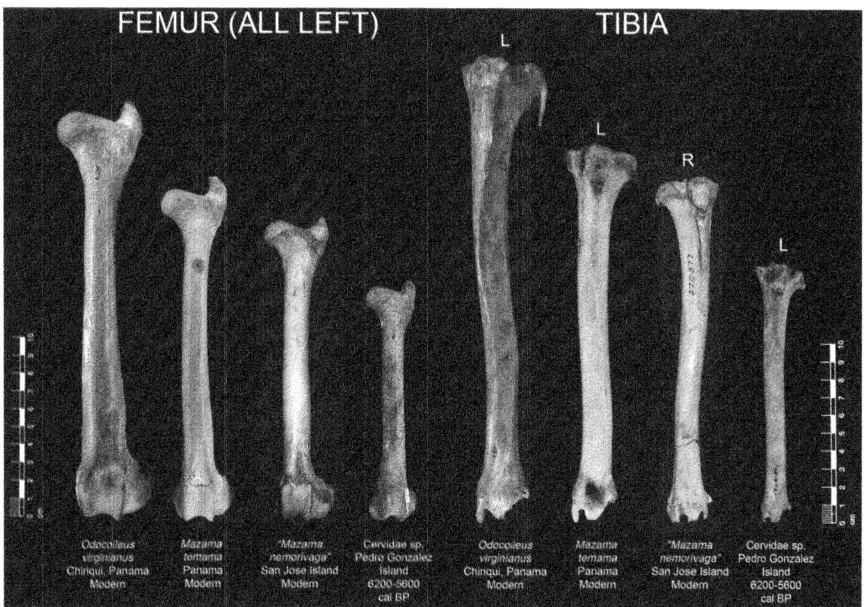

Fig. 10.3 Intact adult femur and tibia from the 6000 cal BP cervid population on Pedro González Island compared with those of modern white-tailed deer, Central American red brocket and a brocket species from San José Island, attributed to *Mazama nemorivaga*

(TL: 143 mm), both adult, are the only long bones that were recovered intact (Fig. 10.3).

10.1.1 Identifying the Source Species

Only three deer species currently occur in Central America outside Mexico. Two are still widespread in appropriate habitats where human population is low or hunting curtailed: the Central American red brocket (*Mazama temama* Kerr 1792) and the white-tailed deer (*Odocoileus virginianus* Zimmerman 1780) (Groves and Grubb 2011). The third extant species is restricted to San José Island in the Pearl Island archipelago in Panama Bay (Pacific) (Fig. 10.1). This small deer was first observed by biologists in the 1940s; Kellogg (1946) assigned it to *Mazama permira*, whereas Handley (1966) treated it as a subspecies of the South American gray brocket (*M. gouazoubira* Fischer 1814), a designation that has been followed by others (e.g., Wilson and Reeder 2005). However, Groves and Grubb (2011) retain the species name *Mazama permira*, in spite of their acknowledged unfamiliarity with specimens. In accordance with more recent phylogeographic data, Rossi et al. (2010) intuitively referred the San José Island population to the Amazonian brown brocket (*Mazama nemorivaga* Cuvier 1817), the remainder of whose distribution lies in northwest South America (Black-Decima et al. 2010). The Amazonian brown brocket and the Amazonian gray brocket are considered by some to be parapatric species (Medellín et al. 1998; Rossi and Duarte 2008). White-tailed deer were introduced to Contadora and San José islands in the Pearl Island archipelago from the mainland after 1970, and thrive there under protection although it is not known how they interact with the native brocket deer on San José Island.

10.1.2 Archaeological Deer Bone Identification

Cervid archaeofaunal samples from pre-Columbian sites on the Pacific side of the Panamanian mainland, which span the period 8000–500 cal BP, appear to contain only white-tailed deer remains (Cooke et al. 2007, 2008). Previously, bone and tooth size rather than morphology have been used to distinguish between *O. virginianus* and *M. "americana"* (now *M. temama* Kerr 1792) at the only pre-Columbian site in Panama at which both taxa have been formally reported (Cerro Brujo, western Caribbean, 1350–1050 cal BP) (Linares and White 1980). Conversely, the presence of *M. "americana"* was rejected by Cooke and Jiménez in samples from several Panamanian pre-Columbian archaeological sites located in ancient wooded savannas and dry littoral zones of the Pacific watershed, in which deer bones were frequent to abundant. This was because none of the adult deer specimens fell within the size ranges of modern Central American red brocket deer (Cooke et al. 2007, 2008). Nor were the typical *Mazama* spike antlers retrieved. However, the tiny size of the PDB Preceramic cervid invalidates a size-based division between *Odocoileus* and *Mazama*, while the demonstrated polyphyly of genus *Mazama* (Duarte et al. 2008) and the need for its re-definition confound the issue even more.

10.1.3 Pedro González Island Archaeological Deer

The most logical inference under the current situation is that the tiny deer encountered by ostensibly the first human seafarers who arrived on Pedro González Island in the Pearl Island Archipelago of Panama Bay ca 6000 cal BP should represent the same phylogenetic lineage as that of the extant population on nearby San José Island, which is now arguably assumed to be an isolated population of the Amazonian brown brocket deer (*Mazama nemorivaga*) (Rossi and Duarte 2008; contra Reid) although being a dwarfed form of the white tailed deer cannot be ruled out due to on-going uncertainties concerning cervid inter-species relationships, which will only be resolved with additional genomic data. However, the extreme anthropogenic modification of the cervid bone sample in the PDB midden, as well as the unusually small size of the deer that produced the specimens, compromised the objective assignation of the remains below the Family level. For this reason, the taxon represented in the archaeological remains was originally reported as an unknown genus and species in Cervidae (Cooke and Jiménez 2009). Specimens of cervids and other mammalian taxa from PDB were evaluated for isotopic analysis in 2012 (Ugan personal communication) but found poor collagen preservation (i.e., <1% yield), lowering the likelihood that quality ancient DNA (aDNA) could be extracted for analysis (Sosa et al. 2013); attempts at aDNA extraction have not proved successful. Therefore, in order to improve understanding of the Pedro González Island sub-fossil deer, we have resorted to protein-based sequence information since it, despite providing a more highly conserved source of molecular sequence information, is more likely to be available in this archaeological material.

10.2 Experimental

10.2.1 Materials for Collagen Analyses

Modern specimens of Central American red brocket deer (*M. temama*) from Panama and white-tailed deer (*O. virginianus*) from Panama and Illinois, USA, were obtained from the Smithsonian Tropical Institute's zooarchaeology skeleton reference collection. Modern specimens of deer identified by the collectors as Amazonian brown brocket (*M. gouazoubira*) were obtained from the von Humboldt Institute, Colombia but it should be noted that these identifications precede the split between *M. gouazoubira* and *M. nemorivaga* (Black-Decima et al. 2010; Rossi et al. 2010). Given their geographic distributions all of these individuals are likely to be the latter. Archaeological specimens for collagen fingerprint analysis from Panama comprised 22 deer bones (Cervidae) from the Preceramic midden at Playa Don Bernardo (PDB).

10.2.2 Morphological Analyses

All measurements taken on deer bones followed (von den Driesch 1976). They were compared with corresponding measurements from (1) the modern skeletons of *O. virginianus* and *M. temama*, (2) two individuals of the San José island *Mazama*, and (3) several bones of white-tailed deer from two Preceramic sites in Central Panama: Cerro Mangote (8000–5000 cal BP) and Zapotal (4300–3300 cal BP).Live weights were estimated on the basis of ten astragali following Purdue (1987) and the breadth of ten femur heads following Reitz and Wing (1999).

10.2.3 Collagen Extraction and Digestion

From each of the specimens sampled, small bone chips (~10 mg) were taken and collagen extracted and analysed following Buckley et al. (2011). In brief, samples were demineralised prior to collagen extraction using 1 mL 0.5 M HCl, overnight at 4 °C. The samples were centrifuged (13,000 × g, 5 min) and the supernatant was discarded. The remaining acid-insoluble pellet from each sample was resuspended using 500 μL of 50 mM ABC (pH 7.4) and gelatinised at 70 °C for 3 h. Subsequent to gelatinisation, the sample was centrifuged (13,000 × g, 15 min), precipitating the ungelatinised protein from the supernatant. The supernatant was then removed for tryptic digestion where 2 μL of 1 μg/μL sequencing-grade trypsin solution was added to the supernatant and incubated at 37 °C for 18 h.

To produce peptide mass fingerprints (PMFs), C18 pipette tips were used to purify (desalt and concentrate the peptides) and fractionate the generated peptide mixture. The pipette tips were firstly equilibrated for sample binding, washing, and elution. This was done with two bed volumes of 50% ACN/0.1% TFA, followed by two bed volumes (100 μL) of 0.1% TFA. Post digestion, the samples were centrifuged (13,000 × g, 15 min) and acidified to 0.1% TFA. The samples were then loaded onto the activated C18 pipette tips by aspirating and dispensing with 10 cycles. The pipette tips were then washed twice with 100 μL 0.1% TFA and a stepped gradient of increasing ACN concentration was applied to the tips to fractionate and elute the peptides (100 μL 10% ACN and 50% ACN with 0.1% TFA). The eluent was aspirated and dispensed ten times, dried using a centrifugal evaporator and re-suspended with 10 μL 0.1% TFA.

10.2.4 MALDI-ToF Mass Spectrometric Analysis

1 μL of the sample solution was spotted onto a Bruker ultraflex 384 target plate, mixed together with 1 μL of α-cyano-4-hydroxycinnamic acid matrix solution (1% in ACN/H2O 1:1 v/v) and dried to air. Each of the collagen digest fractions were analysed by Matrix Assisted Laser Desorption Ionization Time of Flight Mass Spectrometry (MALDI-ToF-MS) in reflectron mode using a Bruker ultraflex II MALDI ToF/ToF mass spectrometer equipped with a Nd:YAG smart beam laser. MALDI-ToF peptide mass fingerprints were acquired over a mass range of m/z 700–3700 using 1000 laser acquisitions. Final mass spectra were externally calibrated against an adjacent spot containing five calibrant peptides. To confirm the homology of peptides between different species, tandem MS (MS/MS) was carried out on selected peptide markers (precursor ion selected with 500 laser acquisitions, up to 4500 laser acquisitions were used for the fragment ions, and argon was used as the collision gas).

10.3 Results

10.3.1 Measurements and Estimated Body Mass

The live weights of the PDB deer were estimated to be 5.77 ± 1.1 kg (range: 3.5–7 kg) and 7.7 ± 0.9 kg (range 6.9–9.4 kg) (Table 10.2). Epiphyseal fusion and tooth eruption sequences show that the PDB sample includes juveniles, sub-adults, adults and mature adults (>5 years). The variations in estimated body mass should therefore reflect age and sex differences among individuals.

Table 10.2 Body mass estimates for the ~6000 cal BP deer on Pedro González Island based on the breadth of the femur head and area of the astragalus

Skeletal element	Reference	N	Mean (kg)	SD	Min.	Max.
Femur	Reitz and Wing (1999)	8	7.70	0.86	6.85	9.38
Astragalus	Purdue (1987)	10	5.77	1.10	3.50	7.00

The above estimates align the size of this sub-fossil cervid with that of the two smallest deer in South America: northern pudu (*Pudu mephistophiles* de Winton 1896): 5–6 kg (Loyola et al. 2010), and Mérida brocket deer (*Mazama bricenii* Thomas 1908): 8–13 kg (Lizcano et al. 2010a). They are below the body mass range estimates for two other small brocket species: dwarf red brocket deer (*M. rufina* Pucheran 1951) (Lizcano et al. 2010b): 10–15 kg, and Brazilian dwarf red brocket deer (*M. nana* Hensel 1872): 10–15 kg (Veltrini-Abril et al. 2010). Three male specimens from a Brazilian population of the Amazonian brown brocket (*M. nemorivaga*) weighed between 14 and 15.5 kg (Viera-Rossi et al. 2010). Small populations in *Odocoileus virginianus* lineages are known on islands. They have estimated body masses above 20 kg, e.g., in the western Florida Keys (*O. virginianus clavium*): 20–35 kg (Miller et al. 2002); on Margarita Island, Venezuela (*O. v. margaritae* or *O. margaritae* (Osgood 1910; Molina and Molinari 1999; Molinari 2007)): 28–31 kg, and on Coiba Island, Panama (*O. v. rothschildi*) (Allen 1904), caveat Olson (2008): about 25 kg.

The neighbouring San José deer appear to be somewhat larger than the sub-fossil deer on Pedro González Island. One measure (total length of the tibia and femur) shows that these San José individuals are intermediate in size between Central American adult red brocket from Panama and the sub-fossil deer from Pedro González (Fig. 10.3, Table 10.3). On the other hand, body mass estimated from the femur head depth of specimen M-22177 was 10 kg whereas the astragalus area (Purdue 1987) of specimen M-27716 gave an estimated body mass of 9.2 kg. The same measures show a much lower body mass for the 6000 year Pedro González deer (5.8 ± 1.1–7.7 ± 0.9 kg) (Table 10.2). These size differences can be explained, firstly, by the fact that the land area of San José Island (44 km^2) is three times that of Pedro González (14 km^2) following Foster's rule and, secondly, by a plausible separation time of 9000-8300 uncalibrated radiocarbon years ago (Martín et al. 2016) between the two island deer populations. It is feasible also that the lack of appropriate mammalian and avian predators on the archipelago in Holocene pre-Columbian and modern times, e.g. felids, canids, mustelids and procyonids, also influenced size reduction in deer because of over-population and competition for restricted resources (Handley 1966; Wolverton et al. 2007; Angehr and Dean 2010; van der Geer et al. 2010; Martínez-Polanco et al. 2015). When deer become isolated on islands, changes occur in morphology, not only a gradual reduction in size, but also a tendency towards hypsodont teeth (high-cusped molars), more massive limbs, a reduction in the length of the autopodium relative to the body, and increases in the numbers of fusions in the foot bones (van der Geer et al. 2010). The PDB cervid samples were studied for these indicators and two cases of fusion

Table 10.3 Body mass estimates for the Pedro González sub-fossil deer

Taxon	Age	Locality	N	Mean	SD	Min.	Max.
Cervidae sp.	6200–5500 cal BP	Isla Pedro González	10	5.8	1.1	3.5	7.0
O. virginianus	8000–3000 cal BP	Central Pacific Panama	9	50.6	8.0	37	66.3
O. virginianus	Modern	Illinois and Chiriqui	2	40.0	4.1	36.9	45.1
M. temama	Modern	Central Panama	2	21.7	4.8	16.9	26.4
M. nemorivaga	Modern	Isla San José	1	9.2			

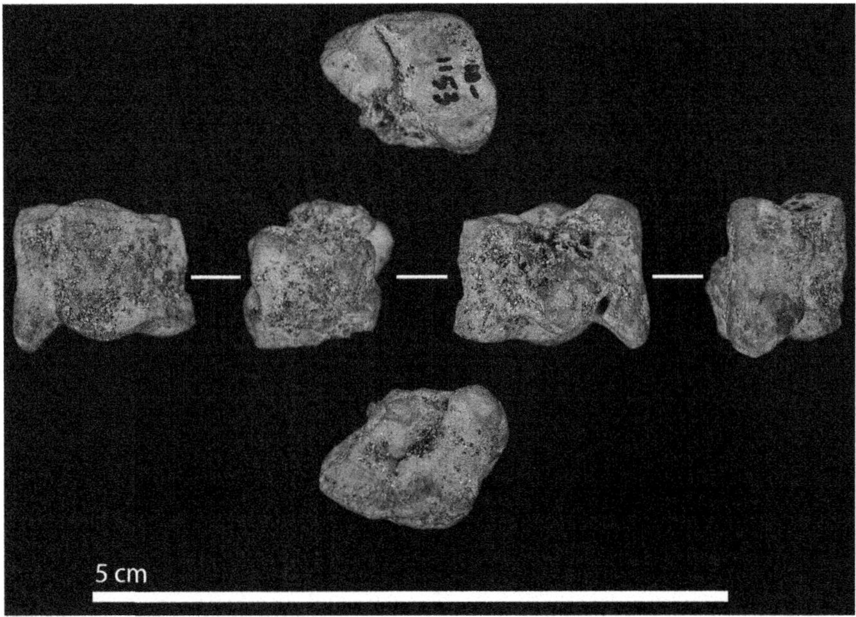

Fig. 10.4 Fused deer podials, Playa Don Bernardo, Isla Pedro González, Pearl Islands: right lunar and cuneiform from Corte 1, Macrostratum III (STRI cat.: 10-1153). *Centre left* to *right*: anterior, lateral, posterior and medial views with proximal view (*top*) and distal view (*bottom*). Photo by Raiza Segundo

in tarsals were observed: (1) a right cuboid fused with the internal cuneiform and (2) a lunar fused with the right cuneiform (Fig. 10.4).

10.3.2 Collagen Fingerprint Analyses

Collagen peptide fingerprints (e.g., Fig. 10.5) could be identified to separate most cervid taxa studied so far at the genus level (to the exception of *Dama* and *Cervus*) but could not be used to separate at the species level. Confounding this matter with

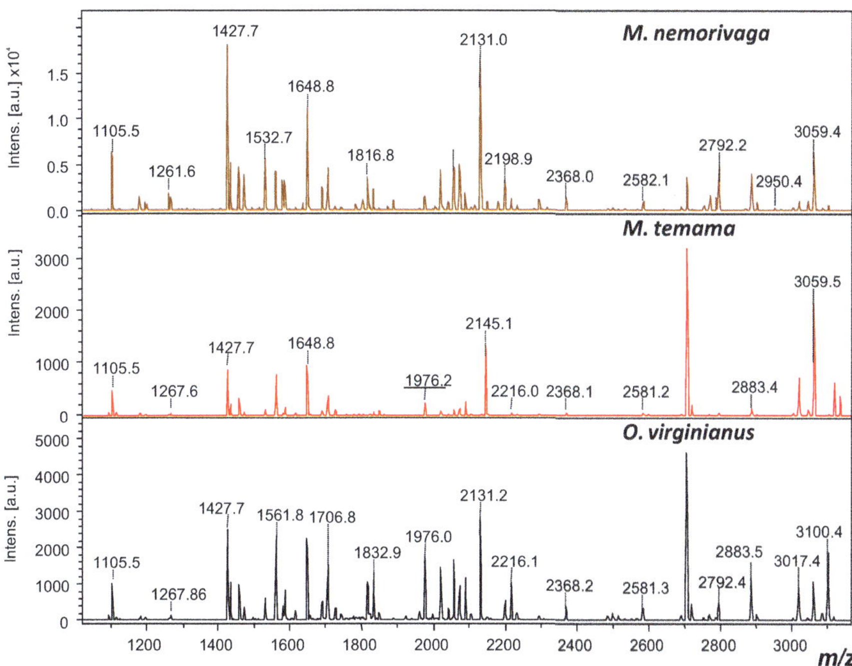

Fig. 10.5 MALDI-ToF-MS peptide mass fingerprints of collagen extracted from *Odocoileus virginianus*, *Mazama temama* and *M. nemorivaga* bone showing only 50% ACN fractions

the current taxa is the demonstrated polyphyly of genus *Mazama* (Duarte et al. 2008), where in this case the two *Mazama* species included in this study, *M. temama* from Panama had distinct markers from the species assumed intuitively by Duarte et al. (2008) to be *M. nemorivaga*, whereas the latter and *O. virginianus* surprisingly shared the same markers.

Collagen fingerprints from modern samples of *M. temama*, deer identified in collections in Argentina as *M. gouazoubira'* and *O. virginianus* (Fig. 10.5) revealed only one of the previously published peptide markers (Buckley et al. 2009; Buckley and Collins 2011; Buckley and Kansa 2011) as being variable: peptide D can be seen at *m/z* 2145.1 in red brocket deer (*M. temama*) but at *m/z* 2131 in white-tailed deer (*O. virgnianus*) and the Amazonian brown brocket deer (*M. nemorivaga*[1]; Fig. 10.5) and all other deer previously analysed (Buckley and Kansa 2011). No other homologous peptide markers could be readily identified that distinguish between these two groups. However, note that all of these South American deer appear to share a previously identified roe deer peptide marker (labelled G in

---

[1]Note that our specimens were listed as *M. gouazoubira*, but sampled from a dataset in which these identifications pre-dated the taxonomic split between gray and brown brocket populations, hence given locality, supported by difference from Argentinian specimens, are assumed as *M. nemorivaga*.

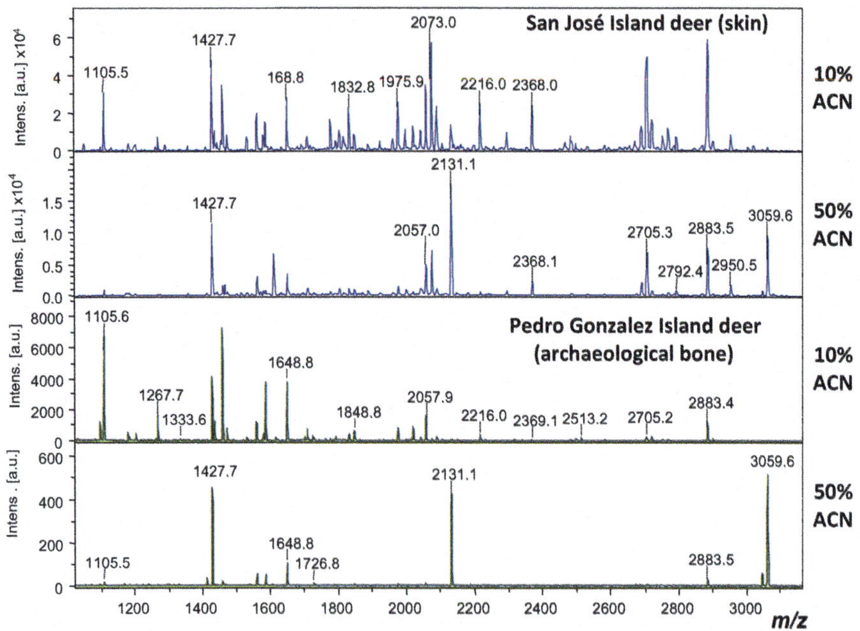

Fig. 10.6 MALDI-ToF-MS peptide mass fingerprints of collagen extracted from the modern San José Island deer skin and the archaeological Pedro González Island deer bone (showing both 10 and 50% ACN fractions)

Buckley et al. 2009) at m/z 3059.4. When compared to these three standards, all archaeological cervids from Playa Don Bernardo as well as the San José deer skin collagen, yielded identical collagen fingerprints with the Amazonian brown brocket and white-tailed deer (e.g., Fig. 10.6; peptide markers needed to separate them from other artiodactyls were C & G—see Buckley et al. 2009).

10.4   Discussion

Although the results of this work clearly indicate that the Pedro González Island deer most likely represent the same species as that present on the nearby San José Island, and that the Central American red brocket (*M. temama*) can be ruled out as an ancestral source, we cannot currently confirm on collagen fingerprints alone whether the ancestral source of these species the white-tailed deer (*O. virginianus*) or the Amazonian brown brocket (*M. nemorivaga*).

10.4.1 The Pearl Islands' Fauna

Although there are no native or introduced deer on Pedro González Island today, a proximal adult femur attributed to *O. virginianus* was found in a shallow (0.4 m) shell-bearing midden (PG-L-106), located on the same beach ridge as PDB (see location in Fig. 10.1) but associated stratigraphically with charred material dated to 1230 ± 40 BP (Martínez-Polanco et al. 2015). The femur was found to be similar in size to adult white-tailed deer femora from pre-Columbian middens in drier areas of Panama's Pacific lowlands and foothills (Cooke et al. 2007, 2008). This is the only cervid specimen that was identified in zooarchaeological samples from test excavations at ceramic sites on ten islands in the Pearl Island archipelago where human occupations between 2300 and 500 calibrated years ago were recorded (Martín et al. 2016).San José Island, the third largest in the Pearl Island archipelago (Panama Bay), is the only location where the Amazonian brown brocket (*M. nemorivaga*) has been reported in Central America. Since its discovery by biologists in the 1940s, this population has been moved from one taxon (*Mazama permira*) to another (*M. gouazoubira*) and yet another (*M. nemorivaga*), but only intuitively apparently following new data on speciation and geographic distribution among South American brocket deer (*Mazama*), but without elucidation of each name change and lacking consensus among specialists. It is apparent from Duarte et al.'s (2008) and Hassanin et al.'s (2012) revisions of South American deer phylogeny, that only molecular sequencing of the San José Island deer population in comparison to different mainland populations of the Central American red brocket (*M. temama*) can be used to resolve satisfactorily the phylogenetic history, not only of this small deer, ostensibly a relic species in disjunction with South American populations assumed to belong to the same lineage, but also of the extinct sub-fossil population on Pedro González Island, which we propose is an earlier form of the same taxon in spite of its separation from the San José Island population for 9000-8300 uncalibrated radiocarbon years (Martín et al. 2016).

10.4.2 Recent Molecular Phylogenies of Mazama and Odocoileus

However, the rapidly advancing molecular research into the phylogeny of New World deer has recently yielded conflicting results. Using cytochrome b sequence data, Pitra et al. (2004) showed white-tailed deer (*O. virginianus*) in a clade with North American mule deer (*O. hemionus*), and sister to *Mazama* sp., whereas Smith et al. (1986), using isozyme data, suggested that the red brocket (*M. americana*) was more closely related to South American forms of white-tailed deer (*O. virginianus*) than to the gray brocket (*M. gouazoubira*). A more recent analysis by Duarte et al. (2008) confirmed that the morphologically-cohesive genus *Mazama* does not form a monophyletic group, but represents separate radiations with high levels of molecular and cytogenetic divergence (Duarte and Merino 1997). Red and gray brockets form two clades with *Odocoileus* spp. nested between two red brocket sub-clades (Duarte et al. 2008). Hassanin et al. (2012) confirmed the existence of two clades for brockets, one containing *M. gouazoubira* and *M. nemorivaga,* and the other, *M. americana* and *M. rufina,* which are themselves grouped together with *Odocoileus.* Several of these associations are in conflict with the results reported here.

Central American brockets were not included in these molecular phylogenetic studies. Since red brocket ("*M. americana*") karyotypes were demonstrated to be distinct between Central and South American populations, Central American red brockets were elevated to species level (*M. temama*) (Jorge and Benirschke 1977; Geist 1998), but they remain poorly studied. *Mazama temama* has not been given a Red List Category and Criterion by the IUCN due to defective knowledge about its distribution in relation to the re-defined South American red brocket (*M. americana*), whereas its persistence and continuity within its known range (Belize to the Panama/Colombia border) are not well documented (Bello et al. 2008). A popular Central American field guide emphasises the unsubstantial nature of the evidence for separating *M. temama* from *M. americana*, and retains the latter species name for *all* red brocket populations (Reid 2009).

10.4.3 Pedro González Island: Past and Present

Available data show that Playa Don Bernardo was abandoned half-way through the sixth millennium cal BP. On this and other islands in the Pearl Island archipelago, no evidence has yet been found for further human occupation until the middle of the third millennium cal BP. The Preceramic settlers of Pedro González hunted the small deer until they abandoned the site, but in diminishing numbers. No cervid bones of this size have been recorded in test excavations conducted at ten ceramic-using sites on five islands. Iguanas (*Iguana iguana*), cane rats (*Zygodontomys*), and spiny rat (*Proechimys*) are the most frequent terrestrial vertebrate taxa by rank order in a sample of only 147 taxonomically significant specimens (Cooke et al. 2016).

Whether it ultimately turns out to be an ultra-dwarfed white-tailed deer with a long history on the archipelago or more likely a lineage of brown brocket closest to *Mazama nemorivaga,* the preservation of the stocks of the San José island deer should be considered a conservation priority. Despite efforts by the owners of the island to protect the deer, poachers still enter clandestinely to hunt the deer and other mammals. The possibility of competition with introduced white-tailed deer and even hybridization is high. On the one hand, obtaining additional tissues for molecular sequencing would improve confidence in the deer's phylogenetic status. On the other hand, archaeological sites in whose middens lies a wealth of information about the Holocene distribution of deer and other vertebrates on the Pearl Island archipelago, are fast disappearing under the machinery of development projects.

10.4.4 The Pedro González Island Terrestrial Environment at 6.2–5.5 cal BP

No archaeobotanical data from sediment cores are available for reconstructing the vegetation on Pedro González Island or elsewhere in the Pearl Island archipelago at the time of first identifiable human occupation. The presence of ecologically healthy forested habitats during the initial part of the Preceramic occupation is strongly suggested by the high rank in the terrestrial vertebrate archaeofauna of green iguana (*Iguana iguana*) and agouti (*Dasyprocta*), as well as *Boa constrictor* and large (>1 m length) colubrid snakes, which, in the absence of mammalian carnivores, were the major predators on these taxa and continue to be so (Cooke et al. 2016). Six of the less frequent terrestrial vertebrate taxa in the PDB archaeofaunas also require heavily vegetated habitats although they are differentially tolerant of forest fragmentation and strong seasonal aridity: paca (*Cuniculus paca*), mud turtles (*Kinosternon leucostomum* and *K. scorpioides*), spiny rat (*Proechimys semispinosus*), rufous tree rat (*Diplomys labilis*) and a medium-sized monkey, probably a capuchin (*Cebus capucinus*). None of these six taxa are currently present on Pedro González Island whereas paca, monkeys and mud turtles have not been recorded anywhere on the Pearl Island archipelago. The presence of mud turtles, which were widely used for food by the pre-Columbian inhabitants of mainland Pacific central Panama (Cooke et al. 2007, 2008), suggests that island streams may have carried more water than today although *Kinosternon* spp. are known to aestivate in the moist mud of water-less streams (Morales-Verdeja and Vogt 1997; Berry and Iverson 2011).

Therefore the pre-human deer on Pedro González likely inhabited vegetation formations similar to the strongly seasonal ones that persist on the Pearl Islands. Small water courses may have contained more water than today. The Preceramic islanders were cultivating maize and other cultigens on the island (Cooke et al. 2016). Therefore they would have cut and burnt the vegetation to prepare fields. How extensive these activities were, however, and whether they would have enhanced the rapid decline of deer and other terrestrial vertebrates that is apparent in the archaeofaunal record by the end of the Preceramic occupation and thereafter, cannot yet be elucidated.

10.5 Conclusions and Future Research

The archaeofaunal deer remains from Playa Don Bernardo hold the key to achieving a far better understanding, not only of the relationship between a mid-Holocene and a modern form of deer possibly in the same lineage, but also of theoretical aspects concerning the effects of dwarfing and isolation on morphology and osteology in artiodactyls. Due to the low likelihood that aDNA extraction and sequencing from these archaeological remains would be feasible, we resorted to a relatively new technique of collagen fingerprinting to identify the pre-Columbian remains. Results indicated that all of the archaeological specimens sampled that yielded collagen fingerprints do not derive from Panamanian *M. temama* (Central American brocket) but from a group of taxa that includes the white-tailed deer and distinctive Colombian populations of *Mazama* whose attribution to species remains in abeyance because of uncertainties regarding the genetics and distribution of *M. gouazoubira* and *M. nemorivaga*. Furthermore, analyses of collagen extracted from the San José deer also match the PDB fingerprint. Future work should try to improve the taxonomic resolution obtainable from protein sequencing methods, perhaps through the study of other non-collagenous proteins (Buckley and Wadsworth 2014) or potentially through screening archaeological samples for aDNA preservation via protein fingerprinting analyses.

Acknowledgements The Pearl Island project (2007–2010) was supported by funds from Panama's Secretaría Nacional de Ciencia, Tecnología e Innovación (SENACYT), the National Geographic Society and the Smithsonian Tropical Research Institute's education programs. Field logistics were coordinated by Reynaldo Tapia and Conrado Tapia. Juan J. Amado, then of the Pearl Island Development Company, and INGEMAR (Panama), generously provided funds for travel, subsistence, archaeology and student participation on Pedro González Island in 2009 and 2010. Test excavations at Playa Don Bernardo benefitted greatly from the assistance of Clara Arango, Eugenia Mellado, Marlene Klages, Yessi Ortiz, Ninel Pleitez, and Aureliano Valencia. Laboratory analyses by MB were supported by funding from the NERC and the Royal Society and carried out in The University of Manchester's Faculty of Life Sciences Biomolecular Analysis Core Research Facility as well as the Manchester Institute of Biotechnology. Special thanks are due to several members of the Pedro González community who assisted in the excavations. This research would not have been possible without *Mazama* samples graciously provided by the Instituto de Investigación de Recursos Biológicos Alexander von Humboldt, Villa de Leyva, Colombia, thanks to the courtesy of Carlos Jaramillo and Claudia Alejandra Uribe Medina. Lastly, thanks are due also de Raiza and Roxana Segundo for assistance with images.

Author Contributions

MB designed and undertook the biochemical research. He planned and wrote the article with RGC. RGC designed and supervised the archaeological field work on the Pearl Island archipelago (2007–2010). FB and JGM directed excavations at the Playa Don Bernardo site and also analyzed material culture. AL participated in the Playa Don Bernardo excavations and assisted with the curation and preliminary analysis of the archaeofaunas. MJ undertook the initial analysis of all the vertebrate remains. MFM updated and expanded the analysis of the cervid remains and organized materials for photography.

References

Allen JA (1904) Mammals from southern Mexico and Central and South America. Order of the Trustees, American Museum of Natural History, New York

Angehr G, Dean J (2010) The birds of Panamá: a field guide. Comstock/Zona Tropical, University of Cornell Press, Ithaca

Bello J, Reyna R, Schipper J (2008) *Mazama temama*. The IUCN Red List of Threatened Species. Version 2014.1. http://www.iucnredlist.org/details/136290/0. Accessed 15 Feb 2016

Berry J, Iverson J (2011) *Kinosternon scorpioides* (Linnaeus 1766) – scorpion mud turtle. Chelonian Res Monogr 5(63):1–15

Black-Decima P, Rossi RV, Vgliotti A, Cartes J, Maffei L, Duarte JMB et al (2010) Brown brocket deer *Mazama gouazoubira* (Fischer 1814). In: Barbanti Duarte JMB, González S (eds) Neotropical cervidology biology and medicine of Latin American deer. Jaboticabal, Brazil and Gland, Switzerland, Funep and IUCN, pp 190–201

Buckley M, Collins MJ (2011) Collagen survival and its use for species identification in Holocene-lower Pleistocene bone fragments from British archaeological and paleontological sites. Antiqua 1(1):e1

Buckley M, Kansa SW (2011) Collagen fingerprinting of archaeological bone and teeth remains from Domuztepe, South Eastern Turkey. Archaeol Anthropol Sci 3(3):271–280

Buckley M, Wadsworth C (2014) Proteome degradation in ancient bone: diagenesis and phylogenetic potential. Palaeogeogr Palaeoclimatol Palaeoecol 416:69–79

Buckley M, Collins M, Thomas-Oates J, Wilson JC (2009) Species identification by analysis of bone collagen using matrix-assisted laser desorption/ionisation time-of-flight mass spectrometry. Rapid Commun Mass Spectrom 23(23):3843–3854

Buckley M, Larkin N, Collins M (2011) Mammoth and mastodon collagen sequences; survival and utility. Geochim Cosmochim Acta 75(7):2007–2016

Cooke R, Jiménez M (2009) Fishing at pre-Hispanic settlements on the Pearl Island archipelago (Panama, Pacific), I: Pedro González Island (4030–3630 cal BCE). In: Makowiecki D, Hamilton-Dyer S, Riddler I, Trzaska-Nartowski N, Makohonienko M (eds) Fishes, culture, environment: through archaeoichthyology, ethnography & history. 15th Meeting of the ICAZ Fish Remains Working Group (FRWG), September 3–9, 2009 in Poznań and Toruń, Poland, vol 7. Poznań, Bogucki Wydawnictwo Naukowe, pp 167–171. Environ Cult

Cooke R, Jiménez M, Ranere A (2007) Influencias humanas sobre la vegetación y fauna de vertebrados de Panamá: Actualización de datos arqueozoológicos y su relación con el paisaje antrópico. Ecología y conservación en Panamá, Smithsonian Tropical Research Institute, Panama City

Cooke R, Jiménez M, Ranere AJ (2008) Archaeozoology, art, documents, and the life assemblage. In: Reitz EJ, Scarry CM, Scudder SJ (eds) Case studies in environmental archaeology. Springer, New York, pp 95–121

Cooke RG, Wake TA, Martínez-Polaco MF, Jiménez-Acosta M, Bustamante F, Holst I, Lara-Kraudy A, Martín JG, Redwood S (2016) Exploitation of dolphins (Cetacea: Delphinidae) at a 6000 yr old preceramic site in the Pearl Island archipelago, Panama. J Archaeol Sci Rep 6:733–756

Duarte JMB, Merino M (1997) Taxonomia e Evolução. In: Duarte J (ed) Biologia e conservação de Cervídeos sul-americanos: Blastocerus, Ozotoceros e Mazama. FUNEP, Jaboticabal, Brazil, pp 1–21

Duarte JMB, González S, Maldonado JE (2008) The surprising evolutionary history of South American deer. Mol Phylogenet Evol 49(1):17–22

Geist V (1998) Deer of the world: their evolution, behaviour, and ecology. Stackpole Books, Mechanicsburg, PA

Groves C, Grubb P (2011) Ungulate taxonomy. John Hopkins University Press, Baltimore

Handley CO Jr (1966) Checklist of the mammals of Panama. In: Wenzel RL, Tipton VJ (eds) Ectoparasites of Panama, vol 861. Field Museum of Natural History, Chicago, pp 753–795

Hassanin A, Delsuc F, Ropiquet A, Hammer C, Jansen van Vuuren B, Matthee C et al (2012) Pattern and timing of diversification of Cetartiodactyla (Mammalia, Laurasiatheria), as revealed by a comprehensive analysis of mitochondrial genomes. C R Biol 335(1):32–50

Jorge W, Benirschke K (1977) Centromeric heterochromatin and G-banding of the red brocket deer, *Mazama Americana temama* (Cervoidea, Artiodactyla) with a probable non-Robertsonian translocation. Cytologia 42:711–721

Kellogg R (1946) The mammals of San José Island, Bay of Panama. Smithson Misc Collect 106 (7):1–4

Linares O, White R (1980) Terrestrial fauna from Cerro Brujo (CA-3) in Bocas del Toro and La Pitahaya (IS-3) in Chiriqui. In: Linares O, Linares A (eds) Adaptive radiations in prehistoric Panama, Peabody Museum monographs, vol 5. Harvard University Press, Cambridge, pp 181–193

Lizcano D, Álvarez S, Delgado-V C (2010a) Dwarf red brocket deer *Mazama rufina* (Pucheran 1951). In: Barbanti Duarte JMB, González S (eds) Neotropical cervidology biology and medicine of Latin American deer. Funep and IUCN, Jaboticabal, Brazil and Gland, Switzerland, pp 177–180

Lizcano D, Yerena E, Alvarez S, Dietrich J (2010b) Mérida brocket deer *Mazama bricenii* (Thomas 1908). In: Barbanti Duarte JMB, González S (eds) Neotropical cervidology biology and medicine of Latin American deer. Funep and IUCN, Jaboticabal, Brazil and Gland, Switzerland, pp 181–184

Loyola L, Escamilo B, Barrio J, Benavides J, Tirira D (2010) Northern pudu. *Pudu mephistophiles* (De Winton 1896). In: Barbanti Duarte JMB, González S (eds) Neotropical cervidology biology and medicine of Latin American deer. Funep and IUCN, Jaboticabal, Brazil and Gland, Switzerland, pp 133–139

Martín J, Bustamante F, Holst I, Lara-Kraudy A, Redwood S, Sánchez-Herrrera L et al (2016) Ocupaciones prehispánicas en Isla Pedro González, Archipiélago de Las Perlas Panamá. Aproximación a una cronología con comentarios sobre las conexiones externas. Lat Am Antiq 27(3):378–396

Martínez-Polanco MF, Jiménez-Acosta M, Buckley M, Cooke RG (2015) Impactos humanos tempranos en fauna ínsula: El caso de los venados de Pedro González (Archipiélago de las Perlas, Panamá). Rev Bioarqueol "Arqueobios" 9(1):202–214

Medellín RA, Gardner AL, Aranda JM (1998) The taxonomic status of the Yucatán brown brocket, *Mazama pandora* (Mammalia: Cervidae). Proc Biol Soc Wash 111(1):1–14

Miller SA, Harley JP, Aloi J, Erickson G (2002) Zoology. McGraw-Hill, New York

Molina M, Molinari J (1999) Taxonomy of Venezuelan white-tailed deer (Odocoileus, Cervidae, Mammalia), based on cranial and mandibular traits. Can J Zool 77(4):632–645

Molinari J (2007) Variación geográfica en los venados de cola blanca (Cervidae, Odocoileus) de Venezuela, con énfasis en *O. margaritae*, la especie enana de la Isla de Margarita. Memoria de la Fundación La Salle de Ciencias Naturales 167:29–72

Morales-Verdeja SA, Vogt RC (1997) Terrestrial movements in relation to aestivation and the annual reproductive cycle of *Kinosternon leucostomum*. Copeia 1997:123–130

Olson S (2008) Falsified data associated with specimens of birds, mammals, and insects from the Veragua Archipelago, Panama, collected by J. H. Batty. Am Mus Novit 3620:1–37

Osgood WH (1910) Mammals from the coast and islands of northern South America. Field Museum Nat Hist, Zool Ser 10:23–32

Pitra C, Fickel J, Meijaard E, Groves C (2004) Evolution and phylogeny of old world deer. Mol Phylogenet Evol 33(3):880–895

Purdue J (1987) Estimation of body weight of white-tailed deer (*Odocoileus virgnianus*) from bone size. J Ethnobiol 71:1–12

Reid F (2009) A field guide to the mammals of Central America and South-east Mexico. Oxford University Press, New York

Reitz EJ, Wing ES (1999) Zooarchaeology. Cambridge University Press, Cambridge

Rossi R, Duarte JMB (2008) Mazama nemorivaga. IUCN Red List of Threatened Species. Version 2011.1. http://www.iucnredlist.org/details/136708/0. Accessed 15 Feb 2016

Rossi RV, Bodmer R, Duarte JMB, Trovati G (2010) Amazonian brown brocket deer Mazama nemorivaga (Cuvier 1817). In: Barbanti Duarte JMB, González S (eds) Neotropical cervidology biology and medicine of Latin American deer. Funep and IUCN, Jaboticabal, Brazil and Gland, Switzerland, pp 202–210

Smith MH, Branan WV, Marchinton RL, Johns PE, Wooten MC (1986) Genetic and morphologic comparisons of red brocket, brown brocket, and white-tailed deer. J Mammal 67:103–111

Sosa C, Vispe E, Núñez C, Baeta M, Casalod Y, Bolea M, Hedges REM, Martínez-Jarreta B (2013) Association between ancient bone preservation and DNA yield: a multidisciplinary approach. Am J Phys Anthropol 151(1):102–109

van der Geer A, Lyras G, de Vos J, Dermitzakis M (2010) Evolution of island mammals: adaptation and extinction of placental mammals on islands. Wiley-Blackwell, Chichester

Veltrini-Abril V, Vogliotti A, Varela D, Barbanti-Duarte J, Cartes J (2010) *Mazama nana* (Hensel 1872). In: Barbanti Duarte JMB, González S (eds) Neotropical cervidology biology and medicine of Latin American deer. Funep and IUCN, Jaboticabal, Brazil and Gland, Switzerland, pp 160–165

von den Driesch A (1976) A guide to the measurement of animal bones from archaeological sites: as developed by the Institutfür Palaeoanatomie, Domestikationsforschung und Geschichte der Tiermedizin of the University of Munich. Harvard University Press and Peabody Museum Press, Cambridge

Wilson DE, Reeder DM (2005) Mammal species of the world: a taxonomic and geographic reference. John Hopkins University Press, Baltimore

Wolverton S, Kennedy J, Cornelius J (2007) A paleozoological perspective on white-tailed deer (*Odocoileus virginianus texana*) population density and body size in central Texas. Environ Manag 39:545–552

Mariana Mondini and A. Sebastián Muñoz

11.1 Introduction and Background

South American camelids have been staple prey in the Andean-Patagonian Neo-tropics ever since humans colonized the region, especially throughout the Holocene, and they have been the subject of domestication processes (Miotti and Salemme 2004; Mengoni Goñalons and Yacobaccio 2006; Bonavia 2008; Wheeler 2012; Borrero 2013, and bibliography therein). They comprise two wild species, the guanaco (*Lama guanicoe*) and the vicuña (*Vicugna vicugna*), as well as two domestic ones, the llama (*Lama glama*) and the alpaca (*Vicugna pacos*), where the llama descended from the guanaco and the alpaca from the vicuña (Wheeler 1995; Kadwell et al. 2001; Marín et al. 2007, among others). Guanacos (80–130 kg) and llamas (80–150 kg) are larger, while vicuñas (35–50 kg) and alpacas (55–65 kg) are smaller (Elkin et al. 1991; Mengoni Goñalons and Yacobaccio 2006, and references therein), all of them bearing low sexual dimorphism (Yacobaccio 2006, 2010; Kaufmann 2009; Kaufmann and L'Heureux 2009; Cartajena 2009).

The osteometry of Neotropical camelids has been the subject of research and discussion over the past century, and it has drawn increasing attention in recent years, given its potential to segregate species, both wild and domestic (for a more detailed review, see Menegaz et al. 1988; Mengoni Goñalons and Yacobaccio

M. Mondini (✉)
Laboratorio de Zooarqueología y Tafonomía de Zonas Áridas (LaZTA), IDACOR, CONICET/
Universidad Nacional de Córdoba, Av. H. Yrigoyen 174, 5000 Córdoba, Argentina

Facultad de Filosofía y Letras, Universidad de Buenos Aires, Ciudad Autónoma de Buenos Aires, Córdoba, Argentina
e-mail: mmondini@conicet.gov.ar

A. Sebastián Muñoz
Laboratorio de Zooarqueología y Tafonomía de Zonas Áridas (LaZTA), IDACOR, CONICET/
Universidad Nacional de Córdoba, Av. H. Yrigoyen 174, 5000 Córdoba, Argentina
e-mail: smunoz@conicet.gov.ar

© Springer International Publishing AG 2017
M. Mondini et al. (eds.), *Zooarchaeology in the Neotropics*,
DOI 10.1007/978-3-319-57328-1_11

2006; Cartajena et al. 2007). Osteometric analyses of camelid archaeological bones supported with modern standards have gained increasing importance ever since their application in the Central Andes by Wing (1972), Kent (1982), and Moore (1989), among others. In the South-Central Andes, our area of interest, they have been applied for the past three decades, especially since the 1990s (Hesse 1982; Elkin et al. 1991; Elkin 1996; Cartajena and Concha 1997; Yacobaccio et al. 1997–1998; Cartajena et al. 2007; Izeta 2007; Mengoni Goñalons 2008, among many others). Yet, not all of the modern standards have been fully published. Moreover, some camelid populations are not as widely represented as others in these standards, potentially impinging some bias on the reference measurement collections. More comparative modern standards are thus required to range as much variation as possible.

Four guanaco subspecies have been proposed on the basis of distribution, size and coloration (González et al. 2006; Bonavia 2008). Recent genetic analyses, though, just separate the northernmost *L. g. cacsilensis* from the remaining guanacos (*L. g. guanicoe*) (Marín et al. 2008; also see Wheeler 2012). The size of the latter has been observed to vary greatly, following a north-south gradient, with much smaller individuals in the South-Central Andes compared to those in the Patagonian populations. However, osteometric data for only a couple of individuals from the former region have been published so far: one from the province of Salta (Elkin et al. 1991; Elkin 1996) and another one from the province of Catamarca (Izeta 2007), both in NW Argentina. Osteometric information on these guanaco populations is thus very scarce; hence, assessing their variability by adding more data is crucial.

The southern vicuña subspecies (*V. v. vicugna*) inhabits NW Argentina (Wheeler 2012; Lichtenstein et al. 2008). Another subspecies of vicuña (*V. v. mensalis*) has been found in the north and west, separated mainly on the basis of size, the latter being smaller. Genetic analyses support this distinction (Marín et al. 2007; Wheeler 2012, and references therein). Although osteometric information on these camelids is relatively more abundant, it is still scarce in terms of intra- and inter-population variability.

Here we present osteometric data of the limb bones of two wild camelid individuals from NW Argentina: a vicuña and a guanaco. As already noticed, the relevance of these new standards is further stressed, on the one hand, by the fact that the vicuña individual is from an area near the southern margins of this species' present range. On the other hand, guanacos from South-Central Andean areas such as this, unlike the larger Patagonian ones, are hardly known, as has often been emphasized (e.g., Mengoni Goñalons and Yacobaccio 2006; Izeta 2007; Cartajena et al. 2007; Mengoni Goñalons 2008; Izeta et al. 2009; Yacobaccio 2010; Gasco and Marsh 2015).

11.2 Materials and Methods

Two individuals have been measured in this study: a 3–4 year old vicuña (*V. vicugna*) male from Antofagasta de la Sierra, at ca. 3800 m.a.s.l., and a 3–4 year old guanaco (*L. guanicoe*) male from Sierra del Aconquija, at >3500 m. a.s.l., both in the province of Catamarca, Argentina. Both individuals belong to the reference collection of the Laboratorio de Zooarqueología y Tafonomía de Zonas Áridas (LaZTA, IDACOR, CONICET-UNC) (LaZTA codes: CMACVVi21 and CMACLGu11, respectively). In order to clean up the bones, the vicuña was macerated in a solution of washing powder and then gently brushed and rinsed, and the guanaco was buried and, after being frozen in a refrigerator for a while, gently brushed and rinsed; none of them was boiled.

The appendicular skeleton was measured following the protocol described by Mengoni Goñalons and Elkin (in Elkin 1996), commonly used in NW Argentina, which in turn has been compared to other commonly used protocols (von den Driesch 1976; Kent 1982). A digital caliper (TESA IP65, resolution = 0.01 mm, accuracy = 0.02 mm) was used for most of the bones, while the most extensive measurements were taken with an osteometric table.

Two measurements of each variable were taken by each observer—the authors, and the average of the four observations was calculated; phalanx measurements were averaged (after Kent 1982), and in the case of first phalanges, fore and rear ones were measured separately. While some archaeological fore and rear first phalanges might be difficult to tell apart, as they differ most ostensibly in their maximum length and the shape of their distal articular end (Kent 1982), recent studies have provided further criteria that apply to other areas such as the usually well-preserved proximal end (Cartajena 2002, 2009; L'Heureux 2008, 2010). Some studies suggest that there is no significant metric difference between front and rear phalanges (Yacobaccio 2010); yet others report significant variation between them, although not necessarily as measured in the same dimensions (Cartajena 2009; L'Heureux 2010, and references therein).

11.3 Results

The camelid measurements are presented in Table 11.1. All of them correspond to right elements, except for the guanaco scapula—as it was broken, the left one was measured instead.

While some measurements are more ambiguous than others in the way they are taken (see von den Driesch 1976), inter-observation variation was ≤3% in all cases, suggesting low inter- and intra-observer variability, in agreement with a previous study (Fraschina et al. 2011).

11.4 Discussion and Conclusions

The measurements presented above are generally consistent with other standards. Some of the most frequently plotted measurements, due to both their usually high representation in the archaeological record and their high discriminatory power for taxa differentiation, are those of the proximal articular surface of the 1st phalanx (Miller 1979; L'Heureux 2010; Gasco et al. 2014). As an example, the articular surface breadth and width (1FA2 and 1FA3K) of the front 1st phalanx are plotted in

Table 11.1 Vicuña and guanaco measurements (in mm) [Measurement definitions after Mengoni Goñalons and Elkin (MG & E, in Elkin 1996), with the corresponding denominations in previous protocols by von den Driesch (1976) (D), Kent (1982) (K), and Menegaz et al. (1988) (M)]

Anatomical parts	Measurements				Camelid species	
	MG & E	D	K	M	Vicuña	Guanaco
Forelimb						
Scapula	ESC1	HS			189.25	205.00
	ESC2		SCA168		173.59	185.06
	ESC3	DHA			189.03	201.32
	ESC4	Ld	SCA167		137.50	153.00
	ESC5	LG	SCA165		27.97	35.13
	ESC6	BG	SCA166		23.73	30.08
Humerus	HUM1	GL	HUM158		205.00	221.00
	HUM2	GLC	HUM159		196.50	211.00
	HUM3	(Bp)	HUM148		47.58	52.22
	HUM4		HUM152		34.96	39.47
	HUM5		HUM153		54.37	61.65
	HUM6	Bd	HUM150		36.36	48.64
	HUM7	(BT)	HUM151		33.56	45.74
Radius-ulna	RUL1	GL	RAUL145		279.00	299.25
	RUL2		RAUL146		246.50	253.25
	RUL3	LO			43.44	49.54
	RUL4	BPC	RAUL134		33.31	44.33
	RUL5	Bd	RAUL141		35.35	43.60
	RUL6	BFd	RAUL144		28.82	38.25
	RUL7		RAUL142		21.48	26.63
Metacarpal	MCP1	GL	MCARP83		197.25	207.75
	MCP2	(Bp)	MCARP59	DTEP	27.01	35.54
	MCP3	Dp	MCARP60	DAPEP	20.90	26.49
	MCP4		MCARP61	DOFAI	19.28	24.49
	MCP5		MCARP63	DOFAE	19.00	22.55
	MCP6	Bd	MCARP77	DTED	33.63	44.67
	MCP7		MCARP78		15.18	21.25
	MCP8		MCARP79		15.08	20.23
	MCP9		MCARP80	DAPED	17.82	22.37
	MCP10		MCARP81		17.93	21.77
1st phalanges	1FA1K ant.		FP1V1		63.60	64.23
	1FA1D ant.	GL		LM	63.35	63.69
	1FA2 ant.	BFp	FP1V2	DTEP	16.00	21.15
	1FA3K ant.		FP1V3		15.09	18.67
	1FA3D ant.	Dp		DAPEP	15.81	19.43
	1FA4 ant.	Bd	FP1V4	DTED	14.48	17.18
	1FA5 ant.		FP1V5	DAEPED	12.60	15.92
	1FA6 ant.			DTD	9.78	12.67
	1FA7 ant.			DAPD	11.13	13.17

(continued)

Table 11.1 (continued)

| Anatomical parts | Measurements | | | | Camelid species | |
	MG & E	D	K	M	Vicuña	Guanaco
Rearlimb						
Femur	FEM1	(GLC)	FEM127		266.50	293.75
	FEM2	GL	FEM128		264.75	292.50
	FEM3	Bp	FEM115		56.59	65.69
	FEM4	DC	FEM117		25.30	28.98
	FEM5	Bd	FEM121		46.94	56.89
	FEM6		FEM123		53.86	61.97
Tibia	TIB1	GL	TIB112		269.50	280.25
	TIB2	Bp	TIB86		52.53	64.18
	TIB3	Bd	TIB102		35.03	44.33
	TIB4	(Db)	TIB103		23.09	27.62
Astragalus	AST1	GLm			31.64	39.25
	AST2	GL1			35.48	44.27
	AST3	Bd			22.39	27.17
	AST4[a]				20.16	26.41
	AST5[b]				26.51	34.44
Calcaneus	CAL1[c]	GL			70.89	81.17
	CAL2	GB			24.00	29.00
	CAL3[a]				31.94	38.96
Metatarsal	MTP1	GL	MTARS44		208.00	210.25
	MTP2	(Bp)	MTARS30	DTEP	26.02	32.26
	MTP3		MTARS31	DAPEP	24.83	28.87
	MTP4		MTARS47		14.53	18.61
	MTP5		MTARS48	DOE	13.95	17.93
	MTP6	Bd	MTARS38	DTED	33.36	43.26
	MTP7		MTARS39		14.63	19.47
	MTP8		MTARS40		15.03	19.55
	MTP9		MTARS41		16.86	20.69
	MTP10		MTARS42		16.98	20.69
1st phalanges	1FA1K post.		BP1V177		57.31	55.72
	1FA1D post.	GL		LM	57.28	55.94
	1FA2 post.	BFp	BP1V178	DTEP	15.61	19.60
	1FA3K post.		BP1V179		13.93	16.24
	1FA3D post.	Dp		DAPEP	14.63	17.56
	1FA4 post.	Bd	BP1V180	DTED	13.51	16.01
	1FA5 post.		BP1V181	DAEPED	12.04	14.56
	1FA6 post.			DTD	9.61	11.34
	1FA7 post.			DAPD	9.97	11.34
2nd phalanges[d]						
	2FA1		P2V6		33.71	32.94
	2FA2[e]				14.29	16.90
	2FA3[f]				10.81	13.68

(continued)

Table 11.1 (continued)

Anatomical parts	Measurements				Camelid species	
	MG & E	D	K	M	Vicuña	Guanaco
	2FA4[g]				12.76	15.47
	2FA5[h]				11.41	13.44

[a]A in Miller (1979)
[b]C in Miller (1979)
[c]X in Miller (1979)
[d]Front and rear second phalanges averaged
[e]Taken as 1FA2
[f]Taken as 1FA3
[g]Taken as 1FA4
[h]Taken as 1FA5

Fig. 11.1 for the cases reported here, and compared with others available in the literature. The new guanaco measurements introduced here are among the smallest ones, and compare to other specimens from NW Argentina. These latter individuals vary in their articular surface breadth and not so much in its width, which looks much more consistent (also see Cartajena 2009). On the other hand, as noticed in most previous studies, some overlap between guanacos and domestic llamas is apparent here, as also seen between vicuñas and domestic alpacas. This includes the cases introduced in this chapter, especially in the case of the guanaco.

On comparing all the measures presented here for the vicuña and the guanaco (Table 11.1), though, variable magnitudes of size difference become apparent. Vicuña measures represent <75% of guanaco ones in the case of HUM6, HUM7, MCP7, and MCP8. Instead, they represent >95% in the case of 1FA1K ant. and 1FA1D ant., and >100%—being vicuña measures larger than guanaco ones—in 1FA1K post., 1FA1D post., and 2FA1. Also, the maximum length of forelimb first phalanges of the smaller vicuña can approach that of the rear first phalanges of the larger guanaco (see also Cartajena 2009; Yacobaccio 2010). Thus, applying the criteria that help distinguish both kinds of phalanges is crucial, as detailed by L'Heureux (2010).

The fact that the forelimb first phalanges of the guanaco and the vicuña measured here are about the same length (1FA1 dimension) differs from what other studies have reported (see Cartajena 2009; Izeta et al. 2009). Yet, guanaco phalanges are much more robust in their diaphyses (1FA1K:1FA6 = 5.08 in guanacos and 6.50 in vicuñas) (see Fig. 11.2). Rear first phalanges are again rather unusual as compared to other studies, vicuña ones being longer than guanaco ones, not only in absolute terms but also as compared to the respective forelimb first phalanges (rearlimb:forelimb 1FA1K = 9.89% in vicuñas and 13.28% in guanacos; see also Moore 1989; Cartajena 2002, 2009). Yet, as in the case of fore first phalanges, guanaco rear ones are more robust than those of the vicuña

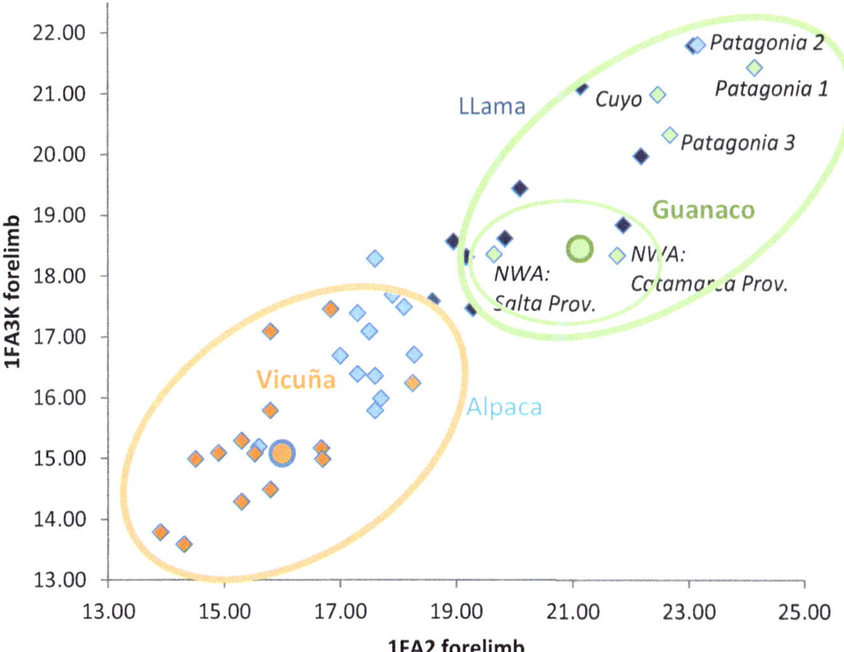

Fig. 11.1 Front 1st phalanx breadth (1FA2) and width (1FA3K) measurements of proximal articular surface. Vicuña (*orange dot*) and guanaco (*green dot*) measured in this study as compared to other published standards (Kent 1982—averages; L'Heureux 2008—average; Izeta et al. 2009; Cartajena 2009; Gasco and Marsh 2015, and references therein): vicuñas in *orange* (*lower oval*), guanacos in *green* (*upper oval*, with NW Argentina individuals in lighter oval), llamas in *blue*, and alpacas in *light blue*. Guanaco geographical origin in italics; *NWA* NW Argentina, *Prov.* Province; Patagonia 1: mean value of guanaco samples measured by Kent (1982, Appendix IV.2); Patagonia 2: mean value of guanaco samples measured by L'Heureux (2008, Table 3.1.2 of Anexo 3—note that this author takes 1FA3D instead of 1FA3K, the former being a bit larger); Patagonia 3: after Izeta et al. (2009)

(1FA1K:1FA6 = 4.93 in guanacos and 5.96 in vicuñas). Vicuña second phalanges are also longer, but thinner, than those of the guanaco.

Thus, at least in some cases, bone gracility may be as important a distinctive trait as gross maximum linear size, an aspect that has been rarely reported (but see Labarca and Prieto 2009). Although diaphyseal thickness is not as commonly considered due to the ambiguity implied in the variable way the caliper can be applied, these measurements were taken in this study after Menegaz et al. (1988) and Mengoni Goñalons and Elkin (in Elkin 1996), and our results suggest that they should be further explored so that they can be eventually linked to specific factors affecting bone robustness.

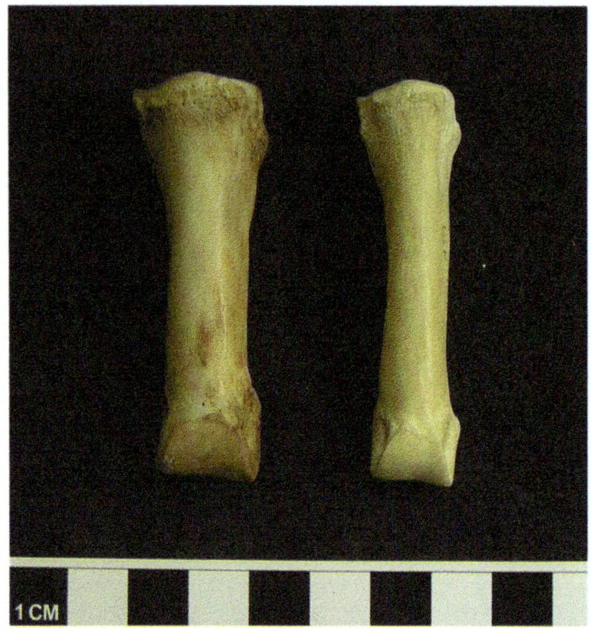

Fig. 11.2 Right front lateral 1st phalanx of the measured guanaco (*left*) and vicuña (*right*)

Results reveal that, despite some metric overlap between both camelid species (see Table 11.1), there are some measures and measure ratios that are more discriminatory regarding taxonomy (such as those plotted in Fig. 11.1), as has often been emphasized (Kent 1982; Miller and Burger 1995, among others). This is especially important since, in addition to the external characteristics that allow distinguishing these camelids (Wheeler 1995; Bonavia 2008), there are only few consistent morphological differences between them at the skeletal level, one of them being in the incisors (Wheeler 1984). Other methods of taxonomic determination of these closely related species are problematic, as in the case of some qualitative morphological traits of the postcranial bones—which vary and overlap among species (e.g., Cartajena 2002)—and ancient biomolecules—which may be hard to recover and identify, especially due to taphonomic and methodological reasons (e.g., Buckley et al. 2014).

This study also contributes to underpinning the assessment of the full size range of the variable guanaco species. The largest Patagonian guanacos were commonly used as a standard for some time (Wing 1972; Miller 1979; Kent 1982; Moore 1989, among others). This suggested a gradient of body size in which this wild species appeared to be typically larger than domestic llamas. While the guanacos from Tierra del Fuego are the largest, Cuyo individuals are intermediate in size within *L. g. guanicoe*, often approaching the Patagonian specimens, as illustrated in Fig. 11.1. The clinal variation in guanaco body size related to latitude, possibly after Bergman's rule (Mengoni Goñalons and Yacobaccio 2006; L'Heureux 2008), probably accounts for this, since Cuyo is a geographically and latitudinally intermediate region between NW Argentina and Patagonia. In fact, the presence of some large archaeological camelid specimens in Cuyo has been interpreted as corresponding to llamas as early as about 5000–4000 years ago (Castro et al. 2013), while llama domestication was apparently only starting in NW Argentina, to the north, where a long-term process of autochthonous llama domestication is being documented (Elkin et al. 1991; Mengoni Goñalons and Yacobaccio 2006). Thus, the possibility that these large Cuyo archaeological bones belong to wild guanacos should not be ruled out (see also Gasco and Cardillo 2014). It is indeed now clear that both wild and domestic *Lama* species range a wider and overlapping size array (Elkin et al. 1991; Mengoni Goñalons and Yacobaccio 2006).

Even though Southern-Central Andean guanacos from NW Argentina and neighboring areas in Chile are much smaller than those in Cuyo and Patagonia, their osteometric values do not seem to overlap with those of the smaller *Vicugna* specimens, as suggested by the case presented here, in addition to other recent studies (e.g., Izeta et al. 2009; Cartajena 2009). Yet, only more robust databases will help discard any potential overlap and eventually allow inferring more confidently their size threshold. Advances in order to address variability are thus crucial, and it should be kept in mind that a few modern osteometric standards cannot be taken as fully representative of the present and past size variation of these native ungulates.

Our study also suggests that there is a wealth of information that can be derived from osteometric studies, not just on taxonomy and domestication. Palaeoclimatic, ecological and evolutionary issues are also of relevance to interpret body size variation in these native camelids, and they have only started to be explored (see for instance L'Heureux 2005, 2007). Further osteometric research will also help improve our understanding of the coevolutionary relationship between these Neotropical native camelids and humans beyond mere taxonomic identification of domestic species (Mengoni Goñalons 2008; L'Heureux 2008; Cartajena 2009, among others).

This all requires the collective building of strong databases, including as much intra- and inter-population variability as possible for all taxa (see Mondini et al. 2013). We are now aware that size and morphological variation in these camelids is great, even greater than genetic variation as detected so far, and possibly not necessarily isomorphic with it. The use of osteometric standards has often been rather essentialist, which was somehow inevitable at early times when there were so few available. Now, instead, this can be avoided by generating new standards and, in so doing, by conceiving size variation of these Neotropical native camelids as potentially great, as a result of multiple conditions and processes.

Several years of individual efforts into bringing to light camelid osteometric standards are certainly paying off, as we are now starting to understand the complexities of metric data in these species across space and time. Continuing efforts will surely help understand the whole range of variation to be expected and, especially, to link it to specific conditions. Hopefully, more solid databases of South American camelid osteometric information will provide insights into a better understanding of this variation, not just as regards species and morphotype identification, but also concerning other biological and ecological aspects of these key Neotropical mammals.

Acknowledgements This research has been funded by PID Res. 1565, categoría A, SeCyT, Universidad Nacional de Córdoba. Dolores Elkin, Alejandra Korstanje and Jorge Mercado kindly helped us get the modern specimens analyzed here. Anahí Ginarte provided the osteometric table. Paola Carmona helped with the photographs. Guillermo Mengoni Goñalons and Lorena L'Heureux shared some bibliography and ideas. Carolina Mosconi helped with the translation into English. Two anonymous reviewers made very useful comments that helped improve this chapter. We are sincerely grateful to all of them, as well as to the 12th ICAZ International Conference Organizers, where this paper was originally presented.

References

Bonavia D (2008) The South American camelids (an expanded and corrected edition). Cotsen Institute of Archaeology Press, University of California, Los Angeles

Borrero LA (2013) Estrategias de caza en Fuego-Patagonia. Comechingonia 17:11–26

Buckley M, Carlini A, Chamberlain A et al (2014) Collagen fingerprinting of vertebrate remains in the Neotropics. In: Libro de Resúmenes, 12th international conference of the International Council for Archaeozoology. FFyH, Universidad Nacional de Córdoba, Córdoba, p 29

Cartajena I (2002) Los conjuntos arqueofaunísticos del Arcaico Temprano en la Puna de Atacama, Norte de Chile. PhD dissertation, Freie Universität Berlin (Printed in microphilm, ABESY Vertriebs GmbH, Gräfelfing, 2003)

Cartajena I (2009) Explorando la variabilidad morfométrica del conjunto de camélidos pequeños durante el Arcaico Tardío y el Formativo Temprano en Quebrada Tulán, norte de Chile. Rev Museo Antropol 2:199–212

Cartajena I, Concha I (1997) Una contribución a la determinación taxonómica de la familia Camelidae en sitios formativos del Loa Medio. Est Atacameños 14:71–83

Cartajena I, Núñez L, Grosjean M (2007) Camelid domestication in the western slope of the Puna de Atacama, Northern Chile. Anthropozoologica 42(2):155–173

Castro S, Gasco A, Lucero G et al (2013) Mid Holocene hunters and herders of Southern Cordillera (Northwestern Argentina). Quat Int 307:96–104

Elkin DC (1996) Arqueozoología de Quebrada Seca 3: indicadores de subsistencia humana temprana en la Puna Meridional Argentina. PhD dissertation, FFyL, Universidad de Buenos Aires

Elkin DC, Madero CM, Mengoni Goñalons GL et al (1991) Avances en el estudio arqueológico de los camélidos en el noroeste argentino. In: Actas de la VII Convención Internacional de Especialistas en Camélidos Sudamericanos, San Salvador de Jujuy, Apr 1991

Fraschina L, Lobbia P, Marozzi A et al (2011) Osteometría de camélidos sudamericanos: evaluando la variabilidad intra e interobservador. In: Álvarez MC, Massigoge A, Izeta AD et al (comps) Libro de Resúmenes. II Congreso Nacional de Zooarqueología Argentina. FCsSs, Universidad Nacional del Centro de la Provincia de Buenos Aires, Olavarría, pp 36–37

Gasco A, Cardillo M (2014) Caracterización morfométrica de la categoría "guanaco andino" (*Lama guanicoe*) en el Centro-Oeste Argentino. Un estudio actual osteométrico como base para análisis zooarqueológicos. In: Libro de Resúmenes, 12th international conference of the International Council for Archaeozoology, FFyH, Universidad Nacional de Córdoba, pp 63–64

Gasco A, Marsh E (2015) Hunting, herding, and caravanning: osteometric identifications of camelid morphotypes at Khonkho Wankane, Bolivia. Int J Osteoarchaeol 25(5):676–689

Gasco A, Marsh E, Kent J (2014) Clarificando variables osteométricas para la primera falange de camélidos sudamericanos. Intersec Antropol 15:131–138

González BA, Palma RE, Zapata B et al (2006) Taxonomic and biogeographical status of guanaco *Lama guanicoe* (Artiodactyla, Camelidae). Mammal Rev 36(2):157–178

Hesse B (1982) Archaeological evidence for camelid exploitation in the Chilean Andes. Säugetierkunde Mitteil 30:201–211

Izeta AD (2007) Zooarqueología del sur de los valles Calchaquíes (Provincias de Catamarca y Tucumán, República Argentina): Análisis de conjuntos faunísticos del primer milenio A.D., B.A.R. international series S1612. Archaeopress, Oxford

Izeta AD, Otaola C, Gasco A (2009) Osteometría de falanges proximales de camélidos sudamericanos modernos. Variabilidad, estándares métricos y su importancia como conjunto comparativo para la interpretación de restos hallados en contextos arqueológicos. Rev Museo Antropol 2:169–180

Kadwell M, Fernández M, Stanley HF et al (2001) Genetic analysis reveals the wild ancestors of the llama and alpaca. Proc R Soc Lond B 268:2575–2584

Kaufmann CA (2009) Estructura de edad y sexo en *Lama guanicoe* (Guanaco). Estudios actualísticos y arqueológicos en Pampa y Patagonia. Sociedad Argentina de Antropología, Buenos Aires

Kaufmann C, L'Heureux L (2009) El dimorfismo sexual en guanacos (*Lama guanicoe*). Una evaluación osteométrica de elementos poscraneales. Rev Museo Antropol 2:182–198

Kent JD (1982) The domestication and exploitation of the South American camelids: methods of analysis and their application to circum-lacustrine archaeological sites in Bolivia and Peru. PhD dissertation, Department of Anthropology, Washington University, University Microfilms International, Ann Arbor (printed in microfilm 1986)

L'Heureux GL (2005) Variación morfométrica en restos óseos de guanaco de sitios arqueológicos de Patagonia Austral Continental y de la Isla Grande de Tierra del Fuego. Magallania 33 (1):81–94

L'Heureux GL (2007) La reducción del tamaño de los guanacos (*Lama guanicoe*) entre el Pleistoceno final y el Holoceno en el extremo austral de Patagonia continental y sus implicancias paleoclimáticas. Archaeofauna 16:173–183

L'Heureux GL (2008) El estudio arqueológico del proceso coevolutivo entre las poblaciones humanas y las poblaciones de guanacos en Patagonia Meridional y Norte de Tierra del Fuego, B.A.R. international series 1751. Archaeopress, Oxford

L'Heureux GL (2010) Estudio biométrico de las primeras falanges de camélidos modernos. Sus implicancias en el análisis de muestras arqueológicas. Rev Werkén 12:109–121

Labarca R, Prieto A (2009) Osteometría de *Vicugna vicugna* Molina, 1782 en el Pleistoceno final de Patagonia meridional chilena: Implicancias paleoecológicas y biogeográficas. Rev Museo Antropol 2:127–140

Lichtenstein G, Baldi R, Villalba L et al (2008) *Vicugna vicugna*. The IUCN Red List of Threatened Species, Version 2014.3. www.iucnredlist.org. Accessed 15 May 2015

Marín JC, Zapata B, González BA et al (2007) Sistemática, taxonomía y domesticación de alpacas y llamas: nueva evidencia cromosómica y molecular. Rev Chilena Hist Nat 80:121–140

Marín JC, Spotorno AE, Gonzalez B et al (2008) Mitochondrial DNA variation, phylogeography and systematics of guanaco (*Lama guanicoe*, Artiodactyla: Camelidae). J Mammal 89 (2):269–281

Menegaz A, Salemme M, Ortiz Jaureguízar E (1988) Una propuesta de sistematización de los caracteres morfométricos de los metapodios y las falanges de camelidae. In: Ratto N, Haber A (eds) De procesos, contextos y otros huesos. Universidad de Buenos Aires, Buenos Aires, pp 53–64

Mengoni Goñalons GL (2008) Camelids in ancient Andean societies: a review of the zooarchaeological evidence. Quat Int 185:59–68

Mengoni Goñalons GL, Yacobaccio HD (2006) The domestication of South American camelids. A view from the South-Central Andes. In: Zeder MA, Bradley DG, Emshwiller E et al (eds) Documenting domestication: new genetic and archaeological paradigms. University of California Press, Berkeley, Los Angeles, pp 228–244

Miller G (1979) An introduction to the ethnoarchaeology of the Andean camelids. PhD dissertation, University of California, Berkeley

Miller GR, Burger RL (1995) Our father the Cayman, our dinner the llama: animal utilization at Chavín de Huantar, Peru. Am Antiq 60(3):421–458

Miotti LL, Salemme MC (2004) Poblamiento, movilidad y territorios entre las sociedades cazadoras-recolectoras de Patagonia. Complutum 15:177–206

Mondini M, Muñoz AS, Fernández PM et al (eds) (2013) Osteometric database of South American camelids (released 2013-10-16). Open Context. http://opencontext.org/projects/0404C6DC-A467-421E-47B8-D68F7090FBCC. Accessed 15 May 2015

Moore K (1989) Hunting and the origin of herding in Peru. PhD dissertation, University of Michigan, University Microfilms International, Ann Arbor (printed in microfilm 1989)

von den Driesch A (1976) A guide to the measurement of animal bones from archaeological sites. Peabody Museum Bull 1:1–136

Wheeler JC (1984) La domesticación de la alpaca (*Lama pacos* L.) y la llama (*Lama glama* L.) y el desarrollo temprano de la ganadería autóctona en los Andes Centrales. Bol Lima 36(6):74–84

Wheeler JC (1995) Evolution and present situation of the South American Camelidae. Biol J Linn Soc 54:271–295

Wheeler JC (2012) South American camelids: past, present and future. J Camelid Sci 5:1–24

Wing E (1972) Utilization of animal resources in the Peruvian Andes. In: Izumi S, Terada K (eds) Andes 4, excavations at Kotosh, Peru, 1963 and 1969. University of Tokyo Press, Tokyo, pp 327–350

Yacobaccio HD (2006) Variables morfométricas de vicuñas (*Vicugna vicugna vicugna*) en Cieneguillas, Jujuy. In: Vilá BL (ed) Investigación, conservación y manejo de vicuñas. Proyecto MACS-Argentina, Buenos Aires, pp 101–112

Yacobaccio HD (2010) Osteometría de llamas (*Lama glama* L.) y sus consecuencias arqueológicas. In: Gutiérrez M, De Nigris M, Fernández P et al (eds) Zooarqueología a principios del siglo XXI. Aportes teóricos, metodológicos y casos de estudio. Ediciones del Espinillo, Buenos Aires, pp 65–75

Yacobaccio H, Madero C, Malmierca M et al (1997–1998) Caza, domesticación y pastoreo de camélidos en la Puna Argentina. Rel Soc Argent Antropol 22–23:389–428

The manufacturer's authorised representative in the EU is Springer
Nature Customer Service Centre GmbH, Europaplatz 3, 69115 Heidelberg,
Germany. If you have any concerns regarding our products, please
contact ProductSafety@springernature.com

Printed and bound by CPI Group (UK) Ltd, Croydon, CR0 4YY

09/01/2026

02031925-0003